REFINING EXPERTISE

Refining Expertise

How Responsible Engineers
Subvert Environmental Justice Challenges

Gwen Ottinger

NEW YORK UNIVERSITY PRESS
New York and London

NEW YORK UNIVERSITY PRESS
New York and London
www.nyupress.org

References to Internet websites (URLs) were accurate at the time of writing.
Neither the author nor New York University Press is responsible for URLs that
may have expired or changed since the manuscript was prepared.

LIBRARY OF CONGRESS CATALOGING-IN-PUBLICATION DATA

Ottinger, Gwen.
Refining expertise : how responsible engineers subvert environmental justice challenges /
Gwen Ottinger.
pages cm
Includes bibliographical references and index.
ISBN 978-0-8147-6237-0 (cl : alk. paper)
ISBN 978-0-8147-6238-7 (pb   )
ISBN 978-0-8147-6239-4 (e-book)
ISBN 978-0-8147-6261-5 (e-book)

1. Petroleum refineries—Environmental aspects—Louisiana—New Sarpy. 2. Environmental
responsibility—United States. 3.   Social responsibility of business—United States. 4.
Petroleum industry and trade—United States. I. Title.

TD195.P4O88 2012

363.738'4—dc23

2012036852

New York University Press books

Manufactured in the United States of America
c 10 9 8 7 6 5 4 3 2 1
p 10 9 8 7 6 5 4 3 2 1

For my parents, Paul and Diane Ottinger

CONTENTS

ACKNOWLEDGMENTS

This project began in earnest the day that Anne Rolfes passed by my bor-rowed desk at Communities for a Better Environment in Oakland, Califor-nia, complimented me on a piece of work that I had done as their science intern, and invited me to visit her organization in Louisiana. It seems fitting, then, that I begin with thanks to her. In the many years that have passed since that first meeting, Anne has offered me lodging, introductions, encour-agement, insight, and an example of the tremendous good that determina-tion combined with optimism can do in the world. For this and more, I am deeply grateful.

Among the wonderful people whom I met through Anne were, of course, the members of Concerned Citizens of New Sarpy and, over time, other resi-dents who had not been involved in the campaign. I thank all of them for welcoming me into their community, opening their homes, and sharing their thoughts on tender subjects; to Myrtle Berteau, Dorothy Gayten, Ida Mitch-ell, Harlon and Janelle Rushing, Audrey Taylor, Gertrude Thompson, Don Winston, and the St. Matthew Baptist Church Women's Ministry, as well as to Iris Carter of Norco, I owe particular thanks for their hospitality, warmth, and friendship. My thanks go also to Tricia Meeks, fellow Louisiana Bucket Brigade volunteer, for her comradeship and thoughtfulness about the dif-ferences between, in her words, "truth seekers" and "rabble rousers"; and to Valerie Kestner, for somehow persuading her understandably skeptical col-leagues to allow me to interview them.

Even beginning this project would not have been possible without Jean Lave and Cathryn Carson, who not only were generous with their time and energy from an early stage of my graduate studies but also had faith in me and what I could accomplish before any of us had any idea where I might be going. To have enjoyed their support and mentorship over all these years has been my great good fortune. I am grateful as well to Gene Rochlin and Jane Summerton for their early (and ongoing) encouragement, and to Paul Rabi-now and Darren Ranco for helping me take the first steps toward framing my fieldwork as a scholarly contribution.

In the writing of the book itself, I benefited from the support of more than a few friends, among them Jody Roberts, Reuben Deumling, Ben Cohen, Ben Gardner, and Kristoffer Whitney, all of whom, at various moments, seemed to understand the project better than I—as did Jane Barnes, who, through a combination of editorial expertise and intelligent unfamiliarity with my particular subject matter, pushed me to see greater significance in my own findings than I might have otherwise. My thanks go to each of them for being willing to engage so thoroughly.

This book, of course, could never have come into being without material support from a variety of sources, for which I am most grateful: the University of California's Center for Information Technology Research in the Interests of Society (CITRIS) funded my initial fieldwork in St. Charles Parish; the John C. Haas Fellowship at the Chemical Heritage Foundation allowed me to make a return trip and spend a postdoctoral year rethinking the work; the Professors as Writers Program at the University of Virginia made it possible for me to hire an editor to help shape the narrative of the book; and a faculty fellowship from the Biological Futures in a Globalized World Initiative at the University of Washington supported the writing of the final chapters. I am particularly indebted to Jody Roberts, who, in his role as manager of the Environmental History and Policy Program at the Chemical Heritage Foundation, created a fellowship opportunity that afforded me the luxury of focusing on writing for an extended period. Finally, my thanks go to Jennifer Hammer of New York University Press, not just for her interest in my manuscript and her helpful feedback as it developed but most of all for giving me the deadlines that, in the end, got the book done.

This book is dedicated to my parents, Paul and Diane Ottinger, who deserve to be acknowledged here, as well. They built for me a foundation of love and support that has made all else possible, and for this I thank them from the bottom of my heart.

ABBREVIATIONS

AMN	Air Monitoring...Norco
BEP	Beneficial Environmental Projects
CAA	Clean Air Act
CAP	Community Advisory Panel
CCN	Concerned Citizens of Norco
CCNS	Concerned Citizens of New Sarpy
CIP	Community Improvement Program
CSR	Corporate Social Responsibility
EPA	Environmental Protection Agency
GNI	Good Neighbor Initiative
LABB	Louisiana Bucket Brigade
LDEQ	Louisiana Department of Environmental Quality
NCA	Norco Civic Association
SCTNA	St. Charles Terrace Neighborhood Association
SEED	Texas Sustainable Energy and Economic Development Coalition
STS	Science and Technology Studies
TELC	Tulane Environmental Law Clinic
VPPP	Voluntary Property Purchase Program

1

The Battlefront

This is the way the world ends
This is the way the world ends
This is the way the world ends
Not with a bang but a whimper.
—T. S. Eliot, "The Hollow Men," 1925

The campaign of Concerned Citizens of New Sarpy against Orion Refining ended with a show of hands in a crowded, windowless, cinder-block room on December 18, 2002.

The campaign had been one of those environmental David-and-Goliath stories about which movies are made. New Sarpy, Louisiana, a working-class town of seventeen hundred people, borders the Orion refinery. The back yards of the modest homes on one side of St. Charles Street end at the refinery's fence; massive storage tanks squat just a few hundred feet away. With the refinery so close, residents were convinced that the toxic chemicals it released into the air were making them sick. So—as in *Erin Brockovich* or *A Civil Action*—the community took on the company, demanding that Orion buy their homes to make it possible for them to relocate to neighborhoods away from industrial pollution. In addition to the usual rallies, press releases, and lawsuits, New Sarpy residents had in their arsenal a novel weapon: the bucket. An inexpensive, homemade air-sampling device, the bucket produced measurements that proved that residents were breathing toxic chemicals released by the refinery. The scientific data supplied by the buckets bolstered both residents' determination to move to a healthier environment and their confidence in their campaign. When in July 2002 Orion offered them money—their choice of home improvement grants or cash payments—to drop an important Clean Air Act lawsuit and continue to live next door, members of Concerned Citizens of New Sarpy (CCNS) angrily denounced the company for trying to buy them off. They vowed to continue their fight for clean air and relocation.

But there will be no movie made about New Sarpy. The bucket will not star as the stone that felled the giant Orion. On that December night in 2002, one week before Christmas, the loosely organized Concerned Citizens group voted to drop their lawsuit and accept a settlement that featured basically the same package of cash payments and home improvement money that Orion had offered—and CCNS had rejected—five months before. In simplest terms, Orion had won. Their money had trumped residents' evidence that they were breathing polluted air.[1]

Or had it? Looking at the campaign in New Sarpy as a familiar story of David versus Goliath, of truth versus power, downplays an important plot twist. On the night of the settlement, leaders of CCNS declared that they had gotten what they wanted all along: clean air. They and Orion officials expressed their mutual appreciation for the respectful conversations through which the settlement had been reached. The corporate Goliath had seemingly become a trusted friend.

The night of December 18, then, marked not only an end to residents' attempts to discredit Orion experts and prove that refinery pollution was harming their health. That night marked the start of a new era of community-industry relations in New Sarpy. It was to be an era of respect, of dialogue, of corporate responsibility.

In the dawning of this new era, there is a movie-worthy story to be told after all—a story of struggle, of resourcefulness, of resilience. It is the story of the experts. It is the story of how petrochemical industry scientists and engineers, and the claims that they made about pollution and health, came under attack from all sides. From residents who disbelieved their reassurances that their plants did no harm. From environmental activists who charged that the industry was harmful on a grand scale. From academics who argued that the experts' truths are not the only, or the best, available. It is the story of how those scientists and engineers resisted those attacks. Of how they drew on important ideas and popular policies to forge a new relationship with residents who mistrusted them. Of how they themselves emerged from the battle changed.

Importantly, it was the newly respectful, cooperative form of community-industry relations—not shows of force—through which petrochemical industry experts regained their status as authorities over technical matters. Far from being a story of the fragility of truth in the face of power, New Sarpy's story is one of the robustness of experts' claims to speak for the truth through clever, fluid alliances with power.

Winning Respect

When the December meeting ended, Jason Carter,[*2] a senior refinery official who had spoken about the settlement plan at the beginning of the meeting, looked pleased to hear of the vote's outcome, which he had awaited in the hallway. A white man[3] in his midforties trained as an engineer, Carter had been frustrated throughout CCNS's campaign by residents' assertions that Orion's unchecked emissions were making them ill.[4] For him, it was indisputable that New Sarpy residents' health complaints were not Orion's fault. Having come to the refinery less than a year after Orion assumed ownership in 1999, Carter conceded that the facility had had a reputation for poor environmental performance and lax safety procedures under its prior owner. He even admitted that, in the start-up process under Orion, the refinery had had a series of flaring incidents that had made it a nuisance to the community. But by the height of residents' campaign in mid-2002, Carter insisted, his refinery had no problems with its emissions. They had been unable to corroborate the results of residents' bucket monitoring, and, moreover, they were working out a settlement with regulators at the Louisiana Department of Environmental Quality to redress the earlier flaring problems and other issues that CCNS had raised in their lawsuit.

Given that the refinery's performance at the time offered no basis for CCNS's continued opposition, Carter attributed the campaign to factors that had nothing to do with science. In particular, he felt that the campaign continued because Orion somehow had not convinced residents that it was "committed to running the place right." He blamed the Louisiana Bucket Brigade (LABB) for this: the involvement of the New Orleans–based environmental health and justice nonprofit, in his view, had turned an early, company-sponsored community meeting into an ambush by irate residents and had subsequently prevented Orion from establishing a relationship with its neighbors. The December 2002 settlement with CCNS indicated that the company had finally been successful in establishing the dialogue with community members that he had sought since arriving at the refinery.

In Carter's account, the turning point in relations between Orion and its neighbors came when he was approached by two CCNS leaders, including Don Winston (also white and of a similar age), who asked if Carter would sit down and talk with CCNS's core leadership. Carter recalled that he quickly agreed, telling the residents that that was just what he had wanted all along. He met with residents without their lawyers or LABB staff, with the stipulation that relocation would not be a subject for discussion. Residents arrived with a list of other demands, which Carter agreed to. Many of these, such as the demand that Orion clean up the industrial trash strewn on a stretch of land just the other side of the fence from New Sarpy, involved issues about which residents felt strongly but of which Orion officials had been unaware—confirming Carter's belief that open lines of communication, not lower emissions, were what was necessary to break the standoff with angry residents.

For Carter, the settlement was a victory—but not a victory of Orion over CCNS, of Goliath over David. Rather, Carter would have called it a victory for both parties. With the campaign behind them, the former antagonists could enjoy a new relationship, characterized by communication and cooperation rather than conflict. The money that residents would receive from the settlement was but one way in which the two groups would work together to improve the community. And, with lines of communication opened, Carter and other refinery officials could better understand and respond to community needs. The new relationship also put an end to spurious (according to Carter) accusations about environmental problems at the refinery and put technical matters back in the hands of experts. Instead of taking bucket samples, residents were asked to report promptly to Orion any noxious odors in the community—the sort that would have triggered bucket monitoring during the campaign—so that refinery staff, committed to safe and environmentally sound operations, could locate and fix the problem.

In contrast to Carter, Anne Rolfes looked grim and deflated as she left the December 18 meeting. A white Louisiana native, the gregarious, indefatigable founder of the Louisiana Bucket Brigade had been CCNS's steadiest source of advice, encouragement, and material resources throughout the two years of their campaign. She thought it undeniable that petrochemical pollution in New Sarpy—and in other so-called fenceline communities adjacent to chemical facilities—caused respiratory ailments and other health problems for residents; that this was not an established scientific fact was, in her view, a result of biased studies and, more importantly, the failure of scientists to collect appropriate data in fenceline communities themselves. The campaign in New Sarpy was an effort to move residents out of harm's way. LABB gave community members the means to collect data to show they were being harmed, providing New Sarpy residents with buckets and helping them conduct a community health study. But Rolfes also encouraged and assisted residents in using traditional organizing strategies, including demonstrations and press conferences, to try to pressure Orion into meeting their demand for relocation.

From Rolfes's perspective, the settlement was a defeat in that it left residents next door to Orion, breathing dangerous chemicals. Moreover, Orion had won the struggle by using blatantly underhanded tactics that ultimately overcame CCNS leaders' resolve to continue their campaign. By offering residents money to drop their lawsuit and remain in New Sarpy, Orion manufactured a split within the community. They then deepened the divisions between CCNS members and previously uninvolved residents who wanted to "take the money" by helping the latter to organize into a rival community group. CCNS leaders were angered by Orion's maneuvering. But, in Rolfes's telling, the refinery's methods eventually made the personal costs of continued resistance too high for CCNS's core group of decision makers, who found themselves plagued by angry recriminations from neighbors and in need of Orion's money to repair hurricane damage to their own homes.

CCNS leaders' decision to settle thus represented to Rolfes a triumph of the oil refinery's sneaky and divisive tactics, of Orion's money and power over residents' evidence and, ultimately, their health. She worried that, by dissolving residents' campaign and taking away the motivations for their air monitoring, the settlement eroded the little power that the residents had gained with respect to the refinery through their organizing and left Orion able, once again, to insist with impunity that health and environmental problems in the neighboring community were not their fault.

Rolfes was not the only one disappointed in the outcome of that December meeting. When the meeting ended, Guy Landry, a white CCNS member in his seventies who had quietly refused to vote for the settlement, went to her and expressed his disgust at his fellow residents' decision to sell out. In doing so, he echoed a complaint that he had made at a press conference months before, when he chastised fellow residents for losing sight of the problem of health in their angry denouncements of Orion.

But the core leaders of CCNS, those who had originally been most critical of Orion and most determined to resist their underhanded efforts to derail the campaign, expressed satisfaction with the meeting's outcome. Don Winston, who had for weeks been bragging that they had finally gotten Orion to sit down and talk "like reasonable businesspeople," explained triumphantly the many ways Orion would be obligated to the community under the settlement. And Ida Mitchell (also white and in her early seventies), although she remained as convinced as anyone that refinery emissions harmed people's health, told me with a defiant look that the community had gotten the clean air that they had wanted all along.[5]

For Mitchell, Winston, and other CCNS leaders, the decision to begin negotiating with Orion was a pragmatic matter. The Louisiana Department of Environmental Quality (LDEQ) had announced in September 2002 that it had reached a settlement with Orion. The LDEQ settlement required the refinery to rectify, and pay penalties for, the violations of the Clean Air Act alleged in CCNS's lawsuit. The enforcement action by the LDEQ made it unlikely that CCNS's suit, filed under a citizen enforcement provision of the act, could go forward. Even if it did, any additional penalties would go to the state treasury and not to the community, whereas community members would benefit directly by dropping the suit and taking Orion's offer of money. Mitchell and others reasoned that, because the LDEQ settlement would guarantee clean air for the community, there was nothing to lose—and much to gain—by settling with Orion. Further, Winston especially emphasized that, if Orion failed to meet its obligations or resumed polluting excessively, there was nothing to stop the community from once again taking action against the refinery.

But while the LDEQ's action marked a victory for clean air, CCNS leaders' sense of triumph on the night of the settlement had as much to do with the way the tone of their interactions with Orion had shifted. Jason Carter and other Orion officials had, throughout the campaign, refused to credit residents' assertions that the refinery was making them sick. Residents regarded this position as blatant dishonesty. They were incredulous when, for example, Orion officials insisted that community members had not been exposed to any hazardous chemicals released during a fourteen-hour fire in a multi-million-gallon gasoline storage tank. Moreover, these and other untruths, in the minds of residents, showed Orion's lack of respect for the community. When a refinery representative suggested that the black sediment that coated their properties was just dirt, residents complained bitterly that Orion treated them as though they could not tell the difference between garden-variety dirt and petrochemical sludge.

Orion's dishonesty and disrespect angered residents and fueled their campaign activities almost as much as their concerns about health effects did. While CCNS's campaign aimed to move community members away from the hazards of the refinery, the air sampling they conducted—during the tank fire, for example—was seen by CCNS members not only to demonstrate that they were being exposed to hazardous chemicals but also, more importantly, to prove that the refinery was lying to them.

Settlement negotiations changed the pattern of disrespect and dishonesty—at least in the minds of the CCNS leaders who participated. Beginning with their first meeting with Carter, CCNS leaders felt that Orion officials had been willing to sit down with residents and talk to them, in Winston's words, "like equals." They began addressing their complaints directly to Carter and other refinery officials instead of organizing meetings and press events around them, and Orion officials, for their part, consistently responded to the issues raised by community leaders.

In their new relationship with refinery managers, CCNS leaders could claim an important victory. They had not won relocation or seen their health concerns acknowledged. But residents' Goliath had nonetheless been brought down to size. With the LDEQ settlement, Orion took responsibility for its environmental effects and ended its disrespectful denials. Key officials had made themselves accessible and accountable directly to the community, as well, through settlement negotiations that would set a precedent for dialogues to follow. In the context of this new relationship, residents accepted Orion's assurances that they could address complaints about flares or smells directly to the company and have refinery scientists and engineers respond seriously to the issues. And even though residents like Guy Landry and Ida Mitchell did not cease to believe that petrochemical emissions affected their health, residents stopped taking bucket samples and publicly challenging now-approachable refinery scientists and engineers about the plant's effects on their health and environment.

* * *

On December 18, 2002, the dominant narrative of community-refinery relations in New Sarpy shifted. During CCNS's campaign, both the community group and its environmentalist supporters, namely Rolfes and LABB, saw residents' interactions with Orion as a battle, a struggle between a powerful, wealthy company and a powerless but determined community. Residents had to fight for acknowledgment of their legitimate health concerns—which they did, with LABB's help, by collecting data on chemical concentrations and illness rates. After the settlement, however, community and refinery saw themselves as partners in dialogue. In their egalitarian communications, residents could raise concerns, including concerns about facility emissions and accidents. Refinery officials took it as their responsibility to inform residents about plant operations and, where residents' concerns pointed to real problems, to identify and address them promptly.

In the shift to the cooperative, communicative model of community-industry relations, refinery scientists and engineers regained their control over scientific claims. As part of the new civility, residents neither collected their own data nor challenged that of experts. But experts, too, altered their approach. Prior to the settlement, experts made pronouncements whose content residents thought implausible and dishonest and whose tone they found disrespectful. After the settlement, experts did not pronounce. Instead, they informed residents of events at the refinery, they listened to concerns, and they took responsibility for flaring and accidents.

If industry scientists' and engineers' technical authority was a central target of CCNS's campaign, the new model of community-industry relations was instrumental to experts' authority in the campaign's wake. The model rested on powerful ideas about what it meant to be a responsible citizen and community member, as well as on public policies that shift responsibility for health, the environment, and social services from central governments to individuals, communities, and corporations. These ideas and policies, characterized by many as "neoliberal,"[6] constituted the cultural terrain on which the campaign in New Sarpy was fought, as I show in this chapter's final section. The same neoliberal cultural terrain shaped the larger battles over expert authority in which New Sarpy's campaign was enmeshed: activist-led fights over the petrochemical industry's claims to sustainability, and academics' struggles to democratize environmental science and policy—to be discussed in turn in the next two sections of the chapter.

Ultimately, it was industry scientists' and engineers' strategic use of the terrain of neoliberal ideas and policies, including the model of community-industry relations to which it gave rise, that enabled them to overcome all three attacks on their technical authority. As the chapters to come will show, the terrain offered important resources that shaped the way scientists and engineers constituted themselves as experts, including the way they presented themselves as authoritative and the basis on which they claimed credibility. This refashioning of expertise, as much as the transformation of community-industry relations itself, discouraged challenges from residents—which, in turn, made it more difficult for other activist and academic detractors to mount their own attacks.

Moreover, neither transformation—of community-industry relations or of expertise—was unique to New Sarpy. In remaking themselves as responsible authorities in a relationship of cooperation and mutual respect with community members, Orion officials adopted an attitude and a set of practices already widespread in the petrochemical industry, including at peer facilities in St. Charles Parish. As a moment of rapid and dramatic change, the end of

CCNS's campaign offers the opportunity to examine the factors that enabled the transformations. That is, the case of New Sarpy shows in microcosm how neoliberal practices and ideologies have allowed not only Orion but companies across the chemical and energy sectors to define their obligations to neighboring communities in a way that both satisfies residents and preserves industry authority over contested environmental and health issues. Providing a window into larger trends, the case further illustrates what environmental activists and engaged academics are up against as they try to push forward criticisms of industry environmental practices and technocratic decision making—criticisms addressed partially, but only partially, in transformations of community-industry relations and petrochemical industry expertise.

Calling to Account

On July 27, 2001, as CCNS's campaign was gathering momentum, a chartered bus pulled up on St. Charles Street, the New Sarpy road nearest Orion's fenceline. Activists—residents fighting refinery neighbors in other communities in Louisiana and Texas and professional organizers from environmental nonprofits—piled out onto the front lawn of CCNS's president, a black woman in her seventies, for a press conference.[7] Clutching protest signs, they surrounded Don Winston as he described a massive tank fire that had occurred at Orion earlier that summer. Among the largest in history, according to Winston, the fire had consumed more than 140,000 barrels of gasoline and burned for over thirteen hours[8]—yet no one from Orion had communicated with New Sarpy residents or tried to get them out of harm's way. Instead, residents were simply told to stay inside with their windows closed and their air conditioning turned off.

With their presence in New Sarpy, environmental activists from around the Southeast were lending support to CCNS's campaign against Orion and their attack on refinery experts. They shared residents' belief that community health was being harmed by petrochemical emissions; they echoed residents' incredulity at Orion scientists' insistence that air quality in the community had not been affected by fumes from the massive fire. But, with protest signs reading "Stop the Bush-Cheney Toxic Two-Step," those environmentalists were fighting their own battle: a battle against changes to the Clean Air Act proposed by the Bush administration (the "toxic two-step"), and a battle, more fundamentally, against public policies that did not take seriously the harm being done to communities by air toxins.

Activists' stop in New Sarpy—part of a multicity "toxic tour" orchestrated by the Texas Sustainable Energy and Economic Development (SEED)

Coalition[9]—dragged CCNS members into this larger struggle. At the national level, a broad coalition of environmental groups was fighting the Bush administration's attempt to eliminate New Source Review, a provision of the Clean Air Act that requires upgrades at refineries and power plants to be reviewed and permitted by regulatory agencies. The environmental groups objected and sought legislative action to block the change. They argued that weakening the Clean Air Act would result in larger quantities of chemicals being released into the air and that the increase in air pollution would have negative public health implications.[10]

Through the toxic tour, New Sarpy became a symbol and a rallying point in the fight to preserve the Clean Air Act. Against the backdrop of the refinery's tanks, activists like Anne Rolfes from LABB and Peter Altman of the SEED coalition (who, like Rolfes, is white) drew on New Sarpy residents' testimony to make their arguments concrete. The tank fire and other pollution problems in the community offered examples of how the Clean Air Act was already failing to protect people's health. Rolling back New Source Review, they argued, could only make things worse for communities like New Sarpy.

Yet New Sarpy and other communities on the toxic tour were more than just illustrations in the national environmental campaign to preserve New Source Review. Community members' alliances with regional and national environmental groups also made them participants in the campaign, as well as in other, far-reaching battles over environmental regulations and corporations' power to evade responsibility for their environmental impacts. By testifying about their experiences of refinery accidents, about the illnesses that they suffered, and about their interactions with dismissive refinery officials, community members helped larger environmental groups to question the scientific claims that underlay the policies that they opposed. Residents' testimony transformed activists' arguments about the effects of petrochemical emissions into concrete, inescapable realities—adding power to environmental groups' critiques by calling experts to account for their impacts.

* * *

The SEED Coalition's toxic tour ended in Washington, D.C., where a group called Clear the Air arranged for residents of refinery-adjacent communities to meet with congressional staffers and ask legislators to weigh in against changes to the Clean Air Act. Residents of New Sarpy were represented in these meetings, as were residents of the neighboring town of Norco. In Norco, which was separated from New Sarpy by the Orion and Motiva refineries and bounded on its other side by Shell Chemical, the African American neighborhood of Diamond was also in the midst of a campaign for relocation—in Diamond's case, away from Shell, whose operating units stood less than fifty feet from some residents' homes.[11]

In Washington, Diamond residents Margie Richard and Jonathon Hawkins described their experiences living next to a petrochemical facility to a member of Senator John Breaux's (D-LA) staff. Richard, a black woman in her sixties, lamented widespread health problems among Diamond's youth, including her ten-year-old grandson. His asthma was so severe, she said, that he had had two near-death experiences in his young life and could not go anywhere without an expensive inhaler and oxygen tank—a picture of which she showed Breaux's staffer. Hawkins, a black teenager, read a poem dramatizing life in Norco: the first-person protagonist plays in the polluted environment around his home, falls victim to a serious illness, and ultimately lies dying in a hospital bed; all the while, he intones, "they [Shell] said it [pollution] wouldn't interfere with me."

In the context of the national policy debate over changes to the Clean Air Act, the testimony of Richard, Hawkins, and other residents illustrated and dramatized environmentalists' arguments that industrial pollution was harmful and would do even more harm if regulations were weakened. In addition, the presence of residents made more compelling environmentalists' admonition that the government should protect the interests of its citizens—residents of communities like Norco and New Sarpy in particular— over the interests of big oil companies.

But the voices of residents, speaking with the authority of experience, were also powerful because they offered environmentalists an additional means of countering the scientific claims of their opponents. When environmentalists like Altman and Rolfes, who also participated in the meetings with staffers, asserted that changes to the Clean Air Act would cause additional harms to the environment and public health, they offered statistics and scientific studies to contradict the statistics and scientific studies of Bush administration and industry officials who favored the changes. In contrast, residents contradicted industry's studies by talking about the illnesses they and their neighbors experienced—illnesses that environmentalists could subsequently refer to as additional evidence of the harms they claimed industrial pollution was doing, and would continue to do, if not properly regulated.

Judged in terms of the scientific studies volleyed back and forth by environmental and industry groups, residents' experience of illness in fenceline communities was relatively weak evidence of industrial facilities' health effects. The experiences of individual residents, especially when recounted by environmental activists, were easily dismissed as mere anecdotes, and apparent clusters of illness readily chalked up to random chance. However, the challenge of residents' testimony to opponents' scientific claims extended beyond their status as evidence, narrowly defined. In narrating their experiences of apparently pollution-related illness in their communities, residents demanded that their illnesses be accounted for. They underscored, moreover, the fact that industry had no satisfactory account to give. Epidemiological studies showing no elevation of disease rates in fenceline communities could not change the *fact* that Richard's grandson, born and raised a stone's throw from a petrochemical plant, could not go anywhere without an inhaler and has had to be hospitalized because he could not breathe. Nor did calling this—and larger patterns of illness in the community—random or isolated go any distance toward explaining why children were sick in the shadow of an industrial facility. Residents' testimony, though inadequate as scientific evidence, represented a challenge to the claims of polluters that could not be fully answered in the terms of industry's scientific studies.

The particular challenge of residents' testimony, it is important to note, could also not be advanced by environmentalists alone. Offered by activists who do not live in fenceline communities, stories of illness among residents are but anecdotal evidence, clearly inferior to quantitative studies. It is only when offered by residents, like Margie Richard, who were themselves living with the effects of pollution that stories of community illness became forceful challenges to those studies. It is Richard who made her grandson's dependence on his inhaler a fact that could not be smoothed into a statistic; it is Richard to whom an explanation was owed. The testimony of residents, then, was a particularly powerful and important part of environmentalists' national-level environmental campaigns because it contested the scientific claims of industry in terms that would not have been available to environmentalists absent their alliances with fenceline communities.

Adding weight to residents' testimony—and the calling-to-account entailed therein—were buckets. Inexpensive, easy-to-use air-sampling devices, the buckets were provided to fenceline communities by the Louisiana Bucket Brigade and other environmental justice groups, which also paid for sample analysis and offered basic technical support. Residents of New Sarpy and Diamond used buckets during their respective campaigns to fill special plastic bags with the air that they breathed and to learn, through laboratory analysis, the levels of toxic chemicals that were in that air.[12] Because air samples are expensive to analyze, residents took samples only when emissions from Orion or Shell, respectively, resulted in particularly noxious odors; nonetheless, over the course of their campaigns, they took several samples, many of which showed high (dangerously high, according to activists) levels of chemicals known to be hazardous to human health. Buckets and bucket results were subsequently incorporated into residents' interactions with the local media, the neighboring facilities, and the environmental regulators responsible for overseeing the industry. After the tank fire at Orion, for example, CCNS announced that residents had measured high levels of carbon disulfide and carbonyl sulfide during the fire; in Norco, a high measured concentration of methyl ethyl ketone arguably caused the US EPA to scrutinize the embattled Shell Chemical facility.[13]

But buckets also became part of residents' testimony in national and international forums and, by amplifying residents' calls-to-account, again helped extend environmentalists' ability to challenge industry science on other-than-scientific terms. In 2001, Margie Richard, as president of Concerned Citizens of Norco, traveled to Holland with a bucket and a bag of air collected from her neighborhood. Her trip to the United Nations Conference on Climate Change, where Shell Chemical was making a presentation, was sponsored by Corporate Watch, a group critical of Shell's environmental practices around the world. Like SEED's toxic tour and Clear the Air's Capitol Hill meetings, the trip was both an opportunity for Richard to call the plight of the Diamond community to the attention of powerful decision makers and a chance for environmental activist groups to draw on Richard's experiences in making a more general case—that, despite its claims to social and environmental responsibility, Shell was culpable for significant environmental degradation and, further, that the whole petrochemical industry and underlying hydrocarbon economy were fundamentally unsustainable.[14]

At the end of Shell's presentation in Holland, Richard had the opportunity to ask a question of the speaker, a senior Shell official.[15] Unscrewing the lid from her bucket and removing the inflated bag inside, she asked if Shell was going to be true to its claims on paper, that it cared about the lives of people and about cleaner air. Richard indicated that there were environmental problems in Norco, where she lived a mere seventeen feet from a Shell chemical plant, and emphasized that she needed to see that something was done about it. She had brought the speaker a gift, she went on: a bag of air from Norco and a sample of the water from the town. The speaker, an elegantly dressed white man with subtly accented English, deflected Richard's question. He said that he was familiar with the issues in Norco but could not address them in that setting. Yet he also refrained from defending the company's environmental record or denying that one of its facilities could be a cause for the health concerns that Richard's testimony implied. Instead, he accepted Richard's bucket sample with the question, "Can I breathe it?"—implicitly acknowledging concerns over the quality of the air and drawing a laugh from his audience.

At the UN conference, the bucket sample from Diamond functioned the same way that Diamond residents' testimony about their illnesses had functioned in meetings with congressional staffers. It rendered concrete and undeniable aspects of residents'—and activists'—claims about local environmental conditions; further, it demanded an account that could not be given in industry's favored scientific terms. Like Richard's asthmatic grandson, the bag of air was an object that could not be made to vanish simply by calling on studies that said that plant emissions did not affect air quality or prevalence of illness in fenceline communities. Bucket samples thus challenged industry studies and scientific claims, but not in scientific terms. Instead, they added to the list of real local impacts for which industry had to account. In fact, by supplementing accounts of illness with documentation of exposure to chemicals that could make people sick, buckets helped residents to weave together pollution and health in residents' testimony—and to demand that industry account for both at once.

By amplifying residents' calls to account, then, bucket sampling heightens the particular challenge residents' testimony makes to industry science. In the context of far-reaching environmental battles—over, for example, the control of toxic chemicals or the sustainability of the petrochemical industry—residents' use of bucket data in their sampling also enhances the ability of environmental groups to challenge scientific claims that deny that industrial pollution is a threat to public health. Just as second-hand accounts of illness in fenceline communities are dismissed as "anecdotal evidence," in scientific debates between environmentalists and industry (in which regulators also participate), bucket results tend to be swept aside as "not representative" of air quality or chemical exposures in fenceline communities.[16]

While data from bucket samples are a weak form of evidence judged in scientific terms, they are potent, as objects that must be accounted for, in the testimony of residents because they call attention to the inadequacy of industry science to account for conditions in fenceline communities. Combined with residents' testimony in the service of far-reaching environmental battles, buckets enhance the ability of environmentalists to criticize industry's claims by allowing them not only to attempt to undermine the scientific studies on which those claims are based but also to ground a critique of industry claims in industry experts' inability to explain the experiences of fenceline community residents. Residents' bucket-informed testimony is especially powerful as part of environmentalists' battles because the kinds of challenges to industry science that they make possible—challenges focused on its adequacy and accountability—are less easily overcome by well-staffed, well-funded industrial research efforts than are challenges made in strictly scientific terms.

* * *

The night of December 18, 2002—the night that marked the end of CCNS's campaign and the transformation of community-industry relations in New Sarpy—had ramifications for the authority of experts that extended beyond the small Louisiana town. The new relationship between the refinery and community members not only restored Orion officials' control over scientific and technical matters. By ending New Sarpy residents' participation in events like the SEED Coalition's toxic tour, events that tied the local struggle to national and international environmental issues, the new relationship also weakened environmental activists' attacks on expert claims that downplayed the dangers of petrochemical pollution. With the cessation of CCNS's struggle, the unique challenge to industry's scientific claims—the calling-to-account—made possible through New Sarpy residents' testimony became unavailable to environmental campaigns. At least insofar as they depended on voices from New Sarpy, environmental groups were left to counter industry's claims on more narrowly defined scientific grounds, where the evidence of environmentalists was more easily contested, and overwhelmed, by better-funded industry scientists.

Knowing Locally

On April 17, 2002, I dragged residents of the obscure Louisiana communities into yet another kind of battle. I described New Sarpy residents' bucket monitoring to a demographically diverse group of Berkeley professors and graduate students intent on saving the planet by, in part, understanding and eliminating obstacles to protective environmental policies. In my presentation, I suggested that community members' use of buckets highlighted not only shortcomings in environmental policy but also weaknesses in the way that policy was made. The data that residents produced through their sampling, I claimed, provided important information about communities' exposures to toxic chemicals during accidents at petrochemical facilities—information that needed to be incorporated into policies to protect public health. But those policies, including regulatory standards for air quality, relied exclusively on knowledge generated by scientists and engineers, using methods accepted by experts but not subject to public scrutiny. Bucket monitoring,

I argued, showed why environmental policy, and the science underlying it, could not be left to experts alone. Both science and policy should be made in more democratic ways.

In challenging the extent of experts' authority on policy issues like air quality, I allied myself with a group of politically engaged scholars who advocate for more public participation in setting environmental policies—scholars who theorize the problems of expert-dominated policy making,[17] who develop innovative approaches to incorporating citizens' voices into technical policy discussions,[18] and who evaluate the successes and failures of government initiatives (prompted, often, by scholarly advice) to include its citizens in environmental policy decisions.[19] By suggesting that bucket-monitoring residents could contribute even to the science underlying environmental policy, I also joined the ranks of scholars who argue that democratic approaches to policy making must involve citizens in setting research agendas, defining research questions, and gathering data.[20] Already combatants in grassroots struggles in their own communities, as well as participants in environmentalists' fights to change environmental policies, New Sarpy and Norco residents became, through my presentation, examples in an academic crusade to democratize environmental science and policy. Their use of buckets, as a counterpoint to regulators' approaches to assessing air quality, added to the rationale for the more participatory approaches being developed by scholars.

My social-scientist colleagues are not alone in believing that democratic participation is necessary to deciding a range of policy issues, including environmental ones. Influential scientists' organizations and government bodies also acknowledge the desirability of public involvement in policy. Where my colleagues diverge from these groups—and hope to make changes to prevailing policy approaches—is in their understanding of how citizens should be involved, and especially how they should be engaged with science. Organizations like the UK Royal Society and the American Association for the Advancement of Science note the technical complexity of the policy issues confronting the public and express concern as to whether citizens possess enough scientific knowledge to intelligently navigate those issues.[21] For these groups, and many of the policy makers that they advise, public participation is overlaid on science: scientific knowledge serves as the common foundation on which democratic debate can be built. Furthering democracy is, accordingly, a problem of fostering "public understanding of science," guaranteeing that the public understands the scientific foundations of policy issues well enough to participate intelligently in democratic discussions.

But the nature of citizen participation is imagined rather differently by the advocates of democratization whose league I (and New Sarpy residents) joined when I made my April 2002 presentation. These social scientists, who study science as a social practice, reject the idea that science can be taken for granted as a stable foundation for decision making.[22] Rather, they point out that science relevant to environmental policy issues is inevitably contested, necessarily uncertain, and inherently value laden.[23] Many of these scholars thus advocate for participation that does not depend on strict demarcations between scientific knowledge and the values, preferences, or opinions of the public but that actually involves citizens in defining the issues that need to be addressed by policy and science.[24] Advocates of public participation also call for democratizing knowledge itself, proposing that citizens' "local knowledge"—of, for example, community environmental conditions—be incorporated into policy discussions and that citizens and scientists collaborate on research about the environmental and health hazards in communities.[25]

Social scientists' calls for democratizing environmental policy turn natural scientists' call for "public understanding of science" on its head.[26] Rather than seeing citizens' grasp of scientific facts as essential to good policy, advocates of democratization stipulate that public policy, and even science itself, are likely to be deficient if the insights of the public are not incorporated. Their argument that citizens' understandings of technical issues, though often divergent from those of scientists, are legitimate and even necessary contributions to policy discussions, depends on examples from communities not unlike New Sarpy—communities engaged with experts around issues of environmental contamination and/or community health; communities whose health and environmental quality were ill served by experts' standard approaches to doing science; communities where residents have themselves been involved in the production of knowledge.[27]

Where scholars have argued that scientific knowledge is inherently political and value laden, empirical research in communities has both made the point concrete and demonstrated the consequences of the finding for real environmental and health problems. For example, in a study of interactions between sheep farmers in northern England and government scientists sent to advise them on what to do with flocks contaminated by radioactive fallout from the 1986 Chernobyl nuclear explosion, Brian Wynne shows how scientists' advice, based on computational models, incorporated their orientation to prediction and control. These unacknowledged values hindered farmers' efforts to deal with the situation when scientists offered with great certainty information that turned out to be wrong; accustomed to making decisions in the face of uncertainty, farmers would have been better served, Wynne suggests, by advice that made the limitations of scientific models clear.[28]

Case studies of communities and their interactions with experts have also shown that science and scientists' ways of knowing routinely neglect community members' specialized knowledge—and that their "local knowledge" is necessary to good science. In the case of the sheep farmers, Wynne argues that government scientists' faulty models could have been improved if scientists had learned from farmers about the specifics of local soil types and grazing behaviors.[29] Other scholars have shown that scientists tended to underestimate the health risks posed by polluted waterways because they significantly underestimated the amount of fish eaten by at-risk communities, especially Native American and ethnic-minority communities; creating more accurate risk assessments has depended heavily on community members' participation in the risk-assessment process.[30] The importance of local knowledge to robust science has likewise been demonstrated in studies of fenceline communities' efforts to gather data about illnesses and environmental exposures,[31] as well as in studies of challenges to the medial research establishment by disease sufferers.[32]

The case of bucket-monitoring New Sarpy residents does similar work for the cause of greater public participation in environmental science and policy making. Used to measure exceptionally high levels of chemicals present during accidents and other unplanned releases from industrial facilities, buckets capture data that are neither gathered by facilities or regulatory agencies nor acknowledged by those experts as important to assessing the potential effects of industrial emissions on community health.[33] The contrast between bucket and experts' monitoring shows how "representativeness," a central value of scientific research but not necessarily a universal goal, makes scientists blind to the potential importance of pollution spikes as a contributor to community health problems. Moreover, bucket monitoring demonstrates the value of residents' "local knowledge"—their knowledge of the symptoms, including itchy eyes, shortness of breath, and nausea, associated with peak periods of pollution.

In the academic crusade for democratization of science and policy, the examples of places like New Sarpy (and Cumbrian sheep farms and Native American reservations) are important. They take a key theoretical justification for democratization—that science is never politically neutral and thus should not be given a privileged position in democratic decision making—and make it concrete. They show what we risk if we allow the values inherent in science to go unexamined: neither sheep farmers nor Louisiana environmental regulators will have the right *kind* of information on which to act.

Case studies of community-expert interactions also offer an instrumental justification for broadening participation in science: absent the insights that community members can offer, scientists cannot produce accurate knowledge on which to base sound policies. Without these examples, scholars' arguments for expanded public participation would rest entirely on claims about the nature of science and moral arguments about what is fair and just in democratic society. With the examples, scholars can add to their arguments accounts of what could happen to the health of actual people and the environment in particular places if expert knowledge is allowed to dominate policy-making processes.

* * *

On October 21, 2002, I joined representatives from Orion and Shell, New Sarpy and Norco, the LDEQ and the U.S. Environmental Protection Agency (EPA), LABB and the SEED Coalition, and a few others, around a large conference table in a meeting room at Tulane University. We gathered that Monday morning to spend a couple of hours discussing the problem of air monitoring in fenceline communities—what was and was not being done, how it could be done better—in the first-ever "Monitoring Roundtable," which I had organized at the behest of Anne Rolfes and Denny Larson, a white community organizer then with the SEED Coalition, in my role as LABB volunteer.[34]

The Monitoring Roundtable was, in a sense, an exercise in the kind of democratization that academics like me argue for in their writings: environmental justice activists sponsored the forum as a way of inserting themselves and community members into expert-dominated processes of air quality monitoring and assessment. Larson and Rolfes had long argued that the information about air quality produced by agency and industry scientists and engineers did not accurately represent the environmental conditions in fenceline communities. The experts, they charged, did not monitor near enough to polluting facilities, soon enough after accidents, or at high enough sensitivities to detect the chemicals that were harming community members' health. Bucket monitoring, in the eyes of activists, corrected these shortcomings. It also created a role for nonscientists in the process of making knowledge about air quality.

The Monitoring Roundtable extended activists' participatory push. It brought a topic usually deliberated by experts alone—how knowledge about air quality ought to be made—into a forum that included community members and their activist allies. In the discussion, Shell representatives touted their Norco monitoring initiative, which included an unprecedented density of monitoring stations but collected data using a protocol scorned by activists. Politely refraining from rolling their eyes, Larson, Rolfes, and Don Winston spoke enthusiastically of the possibilities offered by real-time, fenceline monitors and high-tech hand-held devices. And drawing on my analysis of the reasons agencies and industries monitored the way they did, I suggested that Federal Reference Methods—standards for conducting monitoring— needed to be reconsidered, preferably in consultation with communities.

One small skirmish in the battle to democratize science and policy, the Monitoring Roundtable relied on activism in New Sarpy and places like it. On its own, the Louisiana Bucket Brigade would not have been able to persuade regulators and industry scientists to participate in a public discussion of a topic that they considered their domain. Even with cosponsorship from a professor from the Massachusetts Institute of Technology, whose interest in trying out new, potentially community-friendly monitoring techniques in Louisiana had occasioned the roundtable, it is unlikely that LABB and its graduate student volunteer would have been able to summon experts to deliberations on monitoring strategies.

The experts who attended the roundtable were compelled to the discussion because of community activism. Regulatory agency representatives arguably participated because, in March of 2002, activists petitioned the EPA to revoke the LDEQ's authority to administer the Clean Air Act (CAA), using New Sarpy as an example of the state agency's gross negligence and incompetence. When the EPA subsequently met with representatives of New Sarpy, the monitoring roundtable seemed to appeal to them as a palatable way to mollify activists. LDEQ representatives participated at the suggestion of the EPA and, in turn, recommended that a representative from Orion—subject of a recent enforcement act by the LDEQ under the CAA—also attend. Shell representatives attended, apparently, out of a desire to maintain the company's newly amicable relationship with Concerned Citizens of Norco, established in the wake of the community group's hard-fought campaign.

Other efforts to make science and policy more participatory—interventions studied and staged by scholars like me—have similarly depended on the political agitation of particular communities. AIDS activists angered by the way new drugs were tested, for example, were included in setting rules for clinical trials only after an extended period of activism,[35] and communities that collaborate with regulatory scientists on new approaches to risk assessment are invariably mobilized prior to their participation.[36] Organized around controversial issues such as genetically modified organisms and telecommunications reform, consensus conferences and other regional- or national-level deliberations on science and technology,[37] too, depend on the efforts of social-movement groups to make the issues subjects of public concern.

The struggle to democratize science and policy not only relies on communities like New Sarpy for examples that justify, in concrete terms, the need for participation. For scholars who wish to contribute to, as well as argue for, democratization, they provide the sites and occasions for participatory interventions. Communities like New Sarpy provide sites where people feel that it is pressing to say to scientists and engineers, "you are doing your science badly." They offer occasions where experts, policy makers, or both, for a variety of reasons, feel compelled to listen. Without these places and moments, arguments for greater democracy, even ones with the force of concrete examples and instrumental rationales, would remain confined to the pages of academic journals and university press books. Activists' efforts to challenge scientists' methods and insert themselves into decision making give engaged scholars something to engage.

* * *

While I and other academic researchers were using communities like New Sarpy to question the authority of experts—to argue that expert knowledge should be decentered and the insights of ordinary citizens incorporated into public policy debate—New Sarpy residents laid down their buckets and went back to relying on experts for information about the environmental and health effects of the facility. When, in December 2002, they stopped campaigning and embraced a new kind of relationship with Orion scientists and engineers, they not only stopped contributing to environmental activists' criticisms of experts' claims about pollution and health; they also stopped supporting scholars' challenges to experts' dominance in environmental policy processes. They stopped being an example—except in the past tense— of the harms that experts' unacknowledged values and disdain for local knowledge could cause; they stopped providing moments for experiments in participation.

In fact, New Sarpy residents' new, respectful relationship with refinery experts potentially undermined the argument that ordinary citizens should play more of a role in shaping environmental science and the public policies on which it is based. Residents' apparent satisfaction with expert knowledge in the wake of the settlement reintroduced the possibility, suggested by Orion's Jason Carter in his account of the settlement, that nonscientist community members were simply misinformed about the potential effects of refinery operations. Residents' willingness to abandon their buckets opens them to the charge that their criticisms of the refinery's environmental record were really a smokescreen for other, nontechnical grievances. If New Sarpy residents really did just misunderstand science, if their use of buckets was really just political, if better communication with refinery experts was all that was necessary to resolve the conflict, then greater democracy is not even necessary. Scholars like me are on the wrong track.

Fortifying Expertise

On December 18, 2002, when Orion's scientists and engineers successfully redefined their relationship with New Sarpy residents, they turned back three attacks in one smooth motion. They ended CCNS's bucket monitoring—and residents' claims that refinery experts lied about their emissions. They deprived environmental activists of New Sarpy residents' powerful calls-to-account, forcing them to fight experts on their own, scientific turf. And they weakened academic assaults on expert authority by depriving scholars of an important rationale for and site of expanded participation.

But how did they do it? How did refinery scientists and engineers reestablish themselves as *the* legitimate, credible sources of technical information? And why was forging a new kind of relationship with the community—rather than simply overpowering them—so important to this endeavor?

The answers have everything to do with the terrain on which they fought. The battleground that was New Sarpy was shaped by powerful ideas about what it means to be a responsible person and what makes a nice community. So too was it shaped by public policies that favor voluntary initiatives over command-and-control regulations; private entrepreneurship over public services; and mediated agreements among stakeholders over direct state intervention. For scientists and engineers striving to regain control over technical issues, these ideas and policies became resources. By creatively mobilizing these resources, by taking advantage of the strategic positions that they offered, refinery experts simultaneously reshaped their relationships with residents and refashioned their own authority as experts.[38]

In the course of my fieldwork in New Sarpy, I was surprised at how much Orion officials—and the officials at petrochemical facilities in Norco, whom I would eventually interview as well—seemed to care about maintaining a good relationship with the residents of neighboring communities. I was even more surprised by their rationale. More than one of them told me that they were committed to being "good neighbors" because their "license to operate" came from the community—when, technically, it does not. Orion and other

Louisiana petrochemical facilities are licensed to operate not by residents of neighboring communities but by the LDEQ, which issues operating permits to the facilities.

This counterintuitive idea of *communities* granting companies the right to operate, I came to understand, speaks volumes about the political terrain on which struggles like the campaign in New Sarpy play out. Specifically, it represents one aspect of trends in environmental governance that tend to shift the burdens of environmental protection away from governments and to individuals and markets instead[39]—and that, ultimately, helped industry experts to reestablish their expertise despite criticisms from residents, environmentalists, and academics.

Over the last several decades, the role played by agencies like the LDEQ and the U.S. Environmental Protection Agency in protecting the environment has been changing. Government agencies have been called on to find alternatives to top-down regulations that specify pollution-control technologies that companies must use or emissions levels that they must not exceed. In place of these so-called command and control policies, agencies have been experimenting with approaches that shift responsibility for pollution prevention and control to corporations and the free market.[40] Emissions trading schemes, voluntary programs, and industry-sponsored initiatives all rely on companies themselves to identify and pursue opportunities for emissions reduction. In the chemical industry, the scaling back of government regulation in the 1980s led to the establishment of the Responsible Care program, an initiative of the industry's major trade group (the Chemical Manufacturers' Association, since renamed the American Chemistry Council, or ACC) that requires member companies to implement an environmental management system and strive for continuous improvement in environmental performance.[41]

As regulatory agencies have moved from being the drivers of pollution prevention to being the overseers of or partners in industry- and market-based efforts, they have also sought to shift responsibility for addressing citizens' concerns about pollution to companies. Agencies invite public comment on permitting and other decisions as required by environmental laws; however, where serious conflict between community groups and industrial facilities arises, regulators often express the desire to see community and facility work out the issues themselves. In Louisiana, the LDEQ's Community-Industry Relations group routinely responds to contentious situations in fenceline communities by setting up a Community-Industry Panel to provide warring factions the opportunity to air their disagreements and, through dialogue, to work them out.[42] This regulatory approach echoes the strategy, common throughout the chemical industry, of establishing Community Advisory Panels (CAPs) to address or even forestall community grievances.[43]

Through their preference for dialogic approaches, regulatory agencies shift to companies the responsibility for managing community-industry conflict, participating only as a mediator rather than an enforcer or adjudicator. At the same time, they impose a reciprocal obligation on community members. They demand that aggrieved residents engage in reasonable, respectful, face-to-face discussions with industry representatives—and delegitimize both contentious collective-action strategies and demands for regulatory agencies to actively police the industry. Not only facilities but also community members thus assume some of what had been the regulatory agency's responsibility for ensuring the acceptability of industry's environmental performance.

The shifting of responsibility from environmental regulatory agencies to corporations and individuals echoes broader trends in governance in the United States, as well as in many other nations. Since the 1980s, numerous commentators have noted, the role of government has been redefined by powerful political actors: where the central state's primary obligation had been to provide for its citizens, it is now expected, first and foremost, to guarantee the unfettered functioning of the free market.[44] The underlying expectation that human needs are most efficiently met by corporations in competition with one another has led to policies that devolve responsibility for basic social services to lower levels of government and, perhaps more significantly, to for-profit entities. Public schools, for example, are in some municipalities now run by private corporations, as are many states' prisons. In the environmental realm, policies that privatize natural resources, that set up markets for the right to pollute, and—as in the chemical industry— that remove regulations thought to impede the business decisions of companies are all reflections of an ideology, often dubbed "neoliberalism," that places support for free enterprise at the heart of the government's responsibilities.[45]

As the social functions of the state have been redefined, so too have the responsibilities of its citizens. Theorist Nikolas Rose argues that, under what he terms "advanced liberalism," citizens are expected to take charge of their well-being. They are, for example, assigned responsibility for health and illness: no longer is disease regarded as inevitable and naturally occurring; rather, it is something that citizens can avoid through careful management of their personal risks. More generally, individuals are supposed to be "enterprising," always actively engaged in projects of bettering themselves— projects that involve seeking out, and relying on, the advice of a variety of experts.[46] The assignment of certain kinds of responsibility to citizens, like

changes in the responsibilities of the state, is evident in environmental policies, including, for example, policies that rely on the prudent action of individuals to conserve natural resources.[47]

The redistribution of responsibility among government, private entities,
and individuals has been has been criticized extensively by scholars and
activists alike. They charge that the policies of neoliberalism create a kind of
market rule that undermines participatory democracy,[48] contributes to the
depletion of natural resources and the degradation of environmental quality,[49] and further disadvantages economically and socially marginal groups,[50]
among other ill effects. Yet, in part to ground their criticisms, scholars have
shown how these policies, and the ideologies on which they rest, form an
extensive and uneven cultural terrain[51] on which local politics play out. Studies of local environmental politics, for example, have documented how policies of privatization and deregulation have shaped the possibilities for collective action and social change on issues ranging from ecosystem restoration
to environmental justice.[52] In general, these and other studies show how the
neoliberal cultural terrain tends to close down space for dissenting voices
and critical resistance to capitalist projects; however, they also occasionally
note opportunities for progressive politics to gain a foothold in otherwise
unfriendly territory.[53]

Like so many other environmental struggles, Concerned Citizens of New
Sarpy's campaign and New Sarpy residents' subsequent settlement with
Orion played out on neoliberal cultural terrain. In New Sarpy, the characteristic reassignment of responsibility to individuals and private enterprises
manifested in the Louisiana Department of Environmental Quality's reluctance to intervene in the conflict, which put pressure on CCNS to resolve
their campaign through dialogue. It permeated debates about whether the
community had preceded the refinery or vice versa—debates founded on the
idea that individuals had the ability and obligation to evaluate the risks of
living in New Sarpy before moving there. And the increasing burdens placed
on local governments by neoliberal policies, including the need to attract
private investment, fueled residents' anxiety about protecting an image of
their community as a nice place to live and work amid claims that it was
uninhabitably polluted.

As in other places, the local landscape of neoliberalism in New Sarpy
shaped the outcome of CCNS's campaign. In general, the demands of individual and corporate responsibility acted against residents' attempts at collective action; for example, residents' concern for the quality and image of their
community became a point of leverage for Orion as it sought to force residents to abandon their opposition. But more than just helping to determine

the outcome of the campaign, the neoliberal terrain on which CCNS's battle was fought also shaped the fate of expertise. In the wake of residents' criticisms of experts' knowledge, refinery scientists and engineers were able to reclaim their authority by repositioning themselves on this cultural terrain. The local manifestations of neoliberalism—including the relative absence of the state, the push to community-industry dialogue, the concern for community image, and the obligation of informed residential choices—all provided opportunities for experts to reestablish a credible and authoritative position on technical matters.

Over the next several chapters, I show how petrochemical industry scientists and engineers used the terrain of neoliberalism to overcome critiques of their science and reconstruct their expertise. I demonstrate how discourses of individual responsibility and residential choice provided an opportunity to experts to position themselves as informers of responsible choices (chapter 2); how concerns for community image made previously contested expert knowledge seem restorative in the wake of community campaigns (chapter 3); how community-industry dialogue served as a venue for experts not only to demonstrate their technical knowledge but, more importantly, to express their commitment to environmental quality (chapter 4); and how the association of local facilities with multinational corporations—a facet of neoliberalism's push to global free trade—allowed experts to ground their authority in the moral status of their socially responsible parent companies (chapter 5).

With each of these moves, I argue, scientists and engineers positioned themselves on a decidedly neoliberal terrain in ways that support their claims to authority and shield them from attacks by disgruntled residents—as well as from attacks by allied environmental activists and academic advocates. Moreover, their specific positioning, shared by technical experts throughout the petrochemical industry, helped them recover from the critiques of these groups because it redefined the bases for expert authority. On the cultural terrain of neoliberalism, experts do not merely reassert their authority on the basis of their mastery and infallibility in technical matters. Rather, they find new grounds for claiming expertise, building their authority on claims to be responsible and committed to both their facilities and the communities next door.[54]

Conclusion

When New Sarpy residents settled with Orion in December 2002, I believed that the ethnographic research I was conducting had ceased to be about citizen science and challenges to expert knowledge. It had been transformed, I supposed, into an ethnography of community-industry relations. My contributions to social scientists' struggle to democratize environmental science and policy would have to be made through some other project.

Now I think that I was wrong. With this book, I *am* taking up arms in the struggle to democratize environmental science and policy. Focusing on a case where citizens' challenges to science were thwarted, this book furthers the cause by providing insight into why experts' authority on environmental policy issues is difficult to unsettle. It is not just that petrochemical corporations and their experts are powerful and community groups are weak. It is also that expertise is dynamic: refinery scientists and engineers are able to recraft the bases for their authority. By doing so, they can accommodate some of communities' (and activists' and academics') criticisms and make others irrelevant. And, in the process, they not only draw on contemporary policy trends and the ideologies that underlie them; they also tie their authority more firmly to larger structures of power.

Showing how refinery scientists and engineers use neoliberal ideas and policies as resources in refashioning their authority, this book links the struggle for more participatory environmental policy to battles against neoliberalism being fought by other politically engaged social scientists. These researchers have documented the consequences of neoliberal policies—showing them to be multivalent and contradictory but almost always to the detriment of the world's most vulnerable people—in order to help marginalized communities and international social movements combat the deepening inequalities, erosion of democracy, and degradation of the environment associated with neoliberalism. To the extent that neoliberal policies and ideologies help shield expertise from the criticism of not only academics but also environmental activists and grassroots groups, their struggle is necessarily an aspect of the struggle to democratize environmental policy. Simultaneously, identifying neoliberalism's contributions to expert authority provides further insight into the processes through which neoliberal policies tend to consolidate power, undermine democracy, and contribute to worsening environmental conditions.

In a crowded, windowless, cinder-block room in New Sarpy on December 18, 2002, far more was at stake than whether residents of a small Louisiana town would make a deal with a giant refinery or would continue to try to topple it. Community members' determination to challenge refinery scientists and engineers' "facts" about pollution and health with their own data had made them compelling voices in far-reaching struggles to control the environmental effects of the chemical industry; it had also positioned them as important examples in arguments for replacing expert-dominated environmental policy processes with more broadly participatory approaches. But New Sarpy residents' willingness to establish a newly cooperative, respectful relationship with the refinery—and to defer to refinery experts' authority on matters of pollution and health—makes the town a case study in the ways in which new distributions of responsibility and other "neoliberal" innovations make expertise hard to contest, democratic participation hard to achieve, and environmental contamination hard to avoid.

2

Dangerous Stories

If I'da moved here after they built those tanks, I wouldn'ta said a word about it, it would have been my fault. When I moved here, them tanks and all wasn't there. They wasn't there. And they infringed on my rights.
—Harlon Rushing, New Sarpy resident, April 9, 2003

I really believe that this plant is vital to the national energy. . . . Do I think it's more important than my health? No. But, you know, at this point in time . . . I chose to build my house here and live here, you know. So, you know, one day I'll choose to move.
—Harriet Isaac,* New Sarpy resident, May 20, 2003

As I have been telling you war stories, perhaps you have been trying to decide which side to choose. Is it the people of New Sarpy who deserve your sympathies for all they suffered at the hands of the refinery? Or is the refinery a victim of sensationalist charges, trumped up with the help of environmentalist rabble rousers? If you are not already a partisan, I suspect that there is something that you would like to know in order to help you choose sides: Who was there first, the refinery or the community?

This is a question that I have been asked by nearly every audience to whom I have spoken about New Sarpy. At first I was puzzled by its ubiquity. Why should it matter so much? After months of head scratching, Diona,* a fellow worker at the Louisiana Bucket Brigade (LABB), told me a story that helped me understand. A black resident of a community near Baton Rouge burdened by pollution from a variety of local sources, including an Exxon plastics plant, Diona had recently had a confrontation with a government official[1] about a permitting decision. The official, according to Diona, had denied activists' claims that industrial pollution was making them sick.

Diona suggested that if he did not believe them, he ought to live in their neighborhood for a while and see for himself. He replied that he did not *choose* to live there. Diona recounted this response with harrumphing incredulity, an exclamation point in her voice. Her rejoinder, too obvious to state: as though the community had!

Rare in its pointedness, the exchange, as Diona told it, made clear what was at stake when I was asked, "Who was there first?" The question is one about choice—did residents choose to live next to a refinery?—and, ultimately, about the legitimacy of community action against local industrial facilities. If residents were there first, the logic goes, they have some right to complain about pollution from a neighbor that had not been there when they chose to move in. But if the facility, however noxious, was there first, community action is morally suspect: residents chose to live there knowing, presumably, what they were getting into. Their right to subsequently object to the facility's operations is thus dubious.

The question's underlying logic was at first obscure, I believe, because the connection between moral authority and responsible choice is largely taken for granted in American culture. Smokers who suffer from lung cancer are routinely represented as victims of their own poor choices. So too were holders of zero-down mortgages who lost their homes in 2008's economic crisis—with the consequence that these individuals were judged by many Americans as undeserving of government assistance. But responsible choice is arguably just one quality of the worthy citizen under neoliberalism. In keeping with privatized approaches to governance, social theorist Nikolas Rose (1996b) suggests, neoliberal states promote self-governance, obligating subjects to be not merely responsible choosers but fully "enterprising individuals." In Rose's characterization, the enterprising individual strives for personal fulfillment and takes it as his or her responsibility to achieve that fulfillment through acts of choice. The enterprising individual calls on appropriate experts for information that will guide choices and for help in understanding personal struggles *as* matters of choice, but is, fundamentally, autonomous—able to make and execute appropriate choices without the intervention of the state or other social institutions. In the context of this neoliberal logic, the smoker who gets lung cancer is condemned not just for his irresponsible choice to smoke when the associated health risks are well known but also for his larger failure to pursue the satisfaction that would come from a healthy lifestyle free of tobacco and replete with long walks and leafy greens.

The "who was there first?" question about New Sarpy that I so often encountered sought to gauge whether residents had been responsible, or even enterprising, in their choices—and whether they merited sympathy as a result. Accordingly, the answers offered by New Sarpy residents and petrochemical industry officials referred in various ways to the figures of the responsible chooser and the enterprising individual in their attempts to establish or question the moral authority of residents and the legitimacy of their campaign. Precisely how these figures were mobilized, moreover, shaped industry scientists and engineers' ability to claim expertise over contested issues of environmental health.

During and after Concerned Citizens of New Sarpy's (CCNS's) campaign, residents told two kinds of stories about how they came to live in New Sarpy. In strategic stories, CCNS members and their allies asserted unequivocally the community's prior claim to the area, emphasizing that residents chose to live in a peaceful, rural community relatively free of pollution. They *had* made responsible, informed decisions, only to see their choices taken away by the imposition of the refinery. In contrast, residents' kitchen table stories, told to me away from the heat of the campaign, narrated the complexities and contingencies involved in their decisions to move to New Sarpy—before *and* after the neighboring refinery was built— and acknowledged all they did not, or could not, know about the potential health effects of petrochemical pollution. Kitchen table stories were not concerned with establishing residents' moral authority, as were strategic stories, yet they offered a deeper critique of environmental injustices than simple "we were here first!" assertions by highlighting the structures that landed residents in what they came to believe was an unhealthy proximity to the refinery. Kitchen table stories drew attention in particular to the shifting, incomplete, and necessarily situated nature of the knowledge on which residential choices are supposed to be based—calling into question not just the specific claims about pollution and health made by experts (as strategic stories were wont to do) but the very notion that expert knowledge can guide responsible choices.

Petrochemical industry experts, for their part, met residents' stories and their implied critiques with stories in which the enterprising individual was central. Their enterprising stories refused to acknowledge any structural constraints on residents' choices—in neoliberal ideology, the autonomy of the enterprising person belies structure—allowing them to claim residents' continued presence in New Sarpy as proof that environmental quality did not stand in the way of the healthy, fulfilled life that they, as enterprising individuals, must necessarily be pursuing. Simultaneously, by asserting their own status as enterprising individuals, refinery officials were able to offer their decisions to work in (and in some cases live near) petrochemical facilities as evidence that emissions did no harm.

As part of the neoliberal terrain, the figure of the enterprising individual was one important resource for industry scientists and engineers seeking to maintain their own expert authority in the face of partial and contested information about health effects of chemicals. Attributing enterprising-individual status to residents not only allowed them to gloss structural constraints, including constraints on what could be known about health effects; it also made it hard for residents to critique structures of injustice without jeopardizing their own status as enterprising, autonomous persons. And industry scientists and engineers used the idea to paper over gaps in their own knowledge by wrapping together their claims to technical authority with assertions of their moral authority as enterprising persons.

Strategic Histories

Even months after CCNS's settlement with Orion, Irene Masters,* a black woman in her seventies, was insistent about the order of arrivals in New Sarpy. "When I moved here, they didn't have Orion. It was just an empty field," she told me over coffee at her kitchen table one April morning. From her front windows, we could see Orion's massive storage tanks just across the street. "The plant came and found us here. See, we were here first. And like I said, they moved in on us." While she was recounting how things had changed in her forty-eight years in her home on St. Charles Street, her husband, Harold* (also black), came into the room. He seconded her complaints about the way the neighborhood had developed: it was a good neighborhood when they moved in, he said, not bad like it is now. But when Irene said that it was because the plants were not there back then, Harold contradicted her. Reminding her of the petrochemical facilities that operated, or had operated, within a few miles of their home over the last half-century, Harold claimed, "This here been a plant place for years. That's where most of the people around here be making their money at, around here by those plants."

Harold's and Irene's different ways of telling New Sarpy's history have divergent implications for the moral authority of residents, because they frame differently the choices made by residents when they moved to New Sarpy. Irene's was a strategic history, suggesting that residents chose to live in a community where there was no refinery and that they initially had nothing to fear from industry. Harold's account was a kitchen table history, dangerous in the context of residents' efforts to organize against local industrial facilities. Describing New Sarpy as a "plant place" of long standing, his story suggested that industry was the source of development in the area. Further, it implied that residents settled in New Sarpy for the jobs that the plants offered and thus accepted that their neighborhoods would be plant communities, with all the associated hazards. By subsuming the history of residential development into the history of the industry, Harold's story admitted that residents chose to live in an industrialized area.

Harold's description of New Sarpy as a "plant place" is at least partially accurate: the area surrounding New Sarpy does seem always to have been occupied by industry. The Good Hope refinery, which would eventually become Orion, dates back only to the 1970s; however, parts of its site had been occupied by General American Tank Terminals since 1925, and by the Island salt refinery since the 1910s. Just a few miles east of New Sarpy, the Mexican Petroleum Company, later to become PanAmerican, began building its refinery on the site of the Destrehan plantation in 1914; a few miles west, the New Orleans Refining Company (NORCo.) began operating in 1920 and was purchased by Shell in 1928.[2] Even before the development of the petrochemical industry in the early twentieth century, the area was arguably a "plant place": the region's large petrochemical plants were built on the sites of plantations dedicated to the industrial production of sugar cane and, even earlier, indigo.[3]

However, the long history of industry in the area does not mean that New Sarpy was originally, or only, a "plant place." Residential communities were also always already there; petrochemical companies built plants and towns alongside and on top of them. Some of the plants, including the Island Refinery, General American, and Shell Norco, provided on-site housing for their workers and their families, and the plants unquestionably increased the population of the region. But the town of Good Hope, swallowed up by the Good Hope refinery in the 1970s, had a post office as early as 1922, three years prior to the opening of the 77-acre General American facility.[4] And when the New Orleans Refining Company began building its plant, it did so next to Sellers, an established town with its own post office. The area became officially known as Norco—the name that marks it as a company town—only in 1930, when the Norco post office opened, replacing the one in Sellers.[5]

The dangerous history of New Sarpy as a "plant place" thus considers the town as one of several that developed in tandem with a shifting constellation of industrial facilities over the course of the twentieth century. In contrast, Irene's strategic history, which asserted New Sarpy residents' prior occupancy of the area, deleted the larger contexts of industrial development in St. Charles Parish. Instead, she and other CCNS members focused on the construction and expansion of the Orion refinery as an isolated entity, narrating how the growth of the refinery changed their properties and their neighborhood. By telling New Sarpy's history in this way, they

contrasted the relatively rural setting of their chosen homes to the industrialized landscape produced by the refinery's encroachment—emphasizing that they had not chosen the environment in which they had come to
live.

In her early seventies at the time of CCNS's campaign, Ida Mitchell
remembered what the area was like when she moved to Terrace Street as a
child: "Where the tanks are, all that was forest, if you want to call it that.
And my dad and my brothers cleared all that out." The neighborhood was
still sparsely populated, but many of the residents, white and black, farmed
the land alongside her father, who leased it from the government. Much of
New Sarpy was still agricultural by the time Guy Landry and his wife moved
to a rental home in New Sarpy in 1952; when they built a house on Annex
Street three years later, they bought their lot from a family who had decided
to subdivide and sell their farm. At that time, the land adjacent to the subdivision, where the refinery's tanks now stand, "was nothing but old pasture
land," Landry recalled, "most of it grown up in weeds and everything else."

Over the course of the 1950s and 1960s, the St. Charles Terrace Subdivision filled out significantly: Guy Landry reported that his house was the only
one on the part of Annex Street nearest the river when he moved there in
1955; by 1961, most of the houses that still stand on the block had been built.
Elsewhere in the neighborhood, landowners like Ida Mitchell's and Harold
Masters's parents split up their properties so that their children would have
lots on which to build their own homes.

The land next to the subdivision was still an overgrown field in the 1970s,
when Texas businessman Jack Stanley began to build the Good Hope refinery. Clarice Watson, a black resident, told me that when she moved to her
home—which, like the Masters's, now looks out onto the refinery's tanks—in
1973, she could spot rabbits and other wildlife in the field. But the refinery
was already making its presence felt in New Sarpy by the late 1970s. At the
time, Watson was a cafeteria worker at New Sarpy Elementary School, then
located across the field in Good Hope. After the refinery started operating in
the mid-1970s, she remembered, the school would always keep its buses on
stand-by, in case there was an explosion at the refinery that would force them
to evacuate the school in a hurry. Guy Landry, who worked as a pipe fitter
at the refinery from 1978 until 1982, also commented on the frequency of
the refinery's accidents and recalled numerous occasions on which he would
have to tell his crew to run for safety when it seemed an explosion was immi-
nent. It was, in fact, in the wake of a particularly bad accident that injured
several workers in the late 1970s that the elementary school closed and, ultimately, became the main office building for the refinery.

But although St. Charles Terrace residents were aware of the accident-prone refinery from its inception, they seem not to have initially regarded it as a direct or immediate threat. Recalling the neighborhood even as recently as 1980, residents told me that the refinery was in Good Hope, not close by like it was by the time of the campaign. The construction of the storage tanks in the early 1980s, however, changed residents' sense of remove from the refinery. Suddenly the refinery was on their doorstep—and, importantly, it had moved in without their permission.

According to many New Sarpy residents, the parish had an ordinance that restricted industrial construction within sixteen hundred feet of residential areas. The tanks violated that buffer zone, as Guy Landry confirmed at the time:

> I was working for Jack Stanley then. I was happy to be working in that tank field, I had my people working in that tank field. . . . So one day, about this time, in the afternoon, I went in my truck and I got my tape, my 200-foot tape, and I measured from that last tank, I measured to the property line . . . and I remembered where it was. And that following Saturday I went over and I measured through the church yard and across the street, Mrs. Duhe's yard, I come through her yard, I asked if I could measure through, I measured, I got to the next house facing that street over there, Mr. Berteau, I got the okay from him, I came through there, I came through my brother-in-law's yard which is the house right behind me, I went through this yard, my yard, I went through that, this was an empty lot. Sixteen hundred feet was at exactly the ditch on this side of the street over there. That was sixteen hundred feet.

Despite the fact that the buffer zone encompassed 80 percent of the neighborhood, Landry lamented, "We never could get together, nobody could get together, I don't know, to do anything about it." Others, including Harlon Rushing, a white resident, told me that they did try to do something about it at the time, going to the police jury (the parish government) to protest the construction of the tanks. They blamed parish officials for allowing Stanley to build the tanks over their objections, most suggesting that he had somehow paid off the politicians.

Conclusively placing the residential community in New Sarpy before the refinery's construction, the history of the area offered by CCNS members like Irene Masters, Ida Mitchell, Guy Landy, and Harlon Rushing was strategic in the way it represented residents' choices. Their history suggested that, in moving to New Sarpy, residents chose a peaceful, rural community removed from industrial hazards. Furthermore, it suggested that residents were never given a choice about the refinery's proximity: with the help of a corrupt local government, the facility simply moved in on them. The lynchpin of the strategic story, the abrogation of residents' choices, was what relocation—CCNS's primary campaign goal—would redress.

Resting the legitimacy of CCNS's call for relocation on their prior occupation of New Sarpy, CCNS members' strategic history positioned them as responsible choosers and reinforced the idea of responsible choice as the basis for their campaign's legitimacy. As Harlon Rushing put it, "If I'da moved here after they built those tanks, I wouldn'ta said a word about it, it would have been my fault"—suggesting that, had he chosen to live so near to the refinery in the first place, he would have no right to complain about its adverse consequences because he could have decided to avoid them. It was only because his original choice had been a responsible one, because he could not have accounted for the eventual coming of the refinery in making it, that he felt that he was owed the opportunity to decide to move away if he judged the risks posed by pollution and accidents to be unacceptable.

Kitchen Table Tales

Unlike Rushing, not all of the St. Charles Terrace neighborhood's residents could claim to have moved to New Sarpy prior to the 1970s, when the Good Hope Refinery was first built, or even before the construction of the storage tanks in the early 1980s. Omitted from CCNS's campaign, the stories these residents told about coming to New Sarpy tended not to use the language of responsible choice, or to be particularly concerned with establishing residents' moral authority as responsible choosers. Instead, they portrayed the complex, contingent routes by which residents ended up living near a refinery, their evolving understanding of the risks posed by the facility, and their complicated reasons for staying—or leaving—once they became convinced that it was a hazard.

Like Harold Masters's history of New Sarpy as a "plant place," these kitchen table stories admitted the importance of factors, including economics, family, and simple luck, that opened residents to accusations of not making responsible choices—or, at least, not incorporating health risks into their choices. Yet, as dangerous as these stories were to CCNS members' strategic stories, in drawing attention to the structural constraints on residents' decisions they were also potentially dangerous to the very idea of responsible choice as the basis for the campaign's legitimacy.

Coming to New Sarpy

The complexity of New Sarpy residents' decisions is evident in Audrey Taylor's account of how she came to live in her Annex Street home in 1993. Although Taylor, a retired black woman, was among the St. Charles Terrace residents to move in after the refinery's storage tanks were built, her history in the community extended back to a time before the Good Hope refinery: Taylor grew up in New Sarpy, left to go to high school[6] and college, and returned only after her children were grown and she had retired. She explained her trajectory back to New Sarpy not as a calculated choice so much as a confluence of events.

GO: How did you come to live here?

AT: I lived in New Orleans before I moved away. And my daughter had cancer. So we were, she was accepted at Navy hospital. And she had to go for treatments every two weeks and that meant she was going to stay a week there by herself, and so my husband worked for an insurance company. Lucky enough, they had an office in Virginia . . . so he could ask for a transfer. And we moved to Maryland. And we were blessed because she lived three years after they'd given her like six months to live. And so we were thankful for that. So we stayed there eight years, and when we came back to visit my mother for [her] eightieth birthday, someone was telling my husband they were looking for an agency director, because that's what he's always done, insurance, and they needed a director, and so he went for the interview, and they went back and they wanted him to start the next week. . . . The company that he was working for was folding. . . . He likes the insurance business, so we came on home. Then we had to find a place to stay and my brother says, "Well, nobody's living in that house, we'll fix it up." And so . . .

GO: Because your oldest brother had already died by that point?

AT: No, no, he didn't die until I moved here. We came in '93 and he died in '93. So. That's how we got back home. And I guess it's . . . God works out everything, my mother was sick and someone needed to be here and my brother's wife had cancer and she died first. And we buried her that Wednesday and he had a heart attack and died that Saturday. . . . So, like I said, after that, my mother had gotten older, and I'm the oldest, now, he was the oldest and I was next. I was the oldest girl and he was the oldest boy. So, you know, it's just [slight pause] my responsibility, that's it.

GO: To be here?

AT: To see about family. That's it.

In Taylor's account, a number of factors brought her to live again in the neighborhood where she grew up: her daughter's illness, which took the family away from nearby New Orleans to a different part of the country; the opportunity for her husband to continue to work as an insurance director; the availability of the Annex Street house, still owned by Taylor's family even though it had been unoccupied for six years; and her mother's advancing age.

The range of factors driving Taylor's return to New Sarpy is typical of those represented in other residents' decisions to live in the neighborhood. The combination of family connections and the availability of land, especially, was a frequent theme in residents' accounts of how they came to live in New Sarpy. New Sarpy's oldest residents, in their seventies and eighties at the time of CCNS's campaign, settled there as adults in the late 1950s and 1960s. Although the area was still sparsely populated, a number of them, like Ida Mitchell and Harold and Irene Masters, built homes alongside their siblings on land that their parents already owned. Even families who were entirely new to the area frequently moved from Norco, New Orleans, and other Louisiana towns in clusters, with parents and grown children establishing individual households in close proximity.[7] Guy Landry and his wife, for example, lived adjacent to her sister, who moved to New Sarpy within a few years of the Landrys.

In the 1980s and 1990s, these family clusters helped draw members of younger generations back to New Sarpy. Like Audrey Taylor, Harriet Isaac (who was also black) grew up in New Sarpy, left for college, and worked out of state for several years. She returned in 1986, after separating from her husband, and built a home on the vacant back half of her grandmother's lot. She moved back, she said, for family support, but also because she wanted her children to grow up knowing their grandparents and aunts and uncles and cousins: "I wanted [them] to grow up having a *family* background like I did."

Residents, especially those whose ties to New Sarpy went back decades, seldom made mention of economic factors in talking about their decisions to live in the neighborhood. However, there is reason to suppose that moving to land already owned by family members presented an affordable alternative to buying property in a different community. It seemed that residents occupying family land did not necessarily purchase their properties. While few talked about their specific economic arrangements with their families, it was common for families not to "open secession"—to go through the state to formally transfer property or distribute its value to a person's heirs—after a relative died. As a result, some residents' property was legally owned by a number of brothers and sisters or even an extended group of cousins, and they lived there by virtue of some informal agreement with the family. These residents would at least have paid for upkeep on the properties they occupied, and some, like Harriet Isaac, built their own houses on land that belonged to the family. But either course would presumably have been inexpensive compared to actually buying the land.

Some New Sarpy residents did purchase their properties, even when they moved to family land. Don Winston, who moved in the mid-1990s to a block that had long been occupied by the family of his partner—who, unlike Winston, was black[8]—once mentioned how the two of them had had to negotiate with numerous members of the family to ensure that they would legally own the property. Again, residents almost never talked about these transactions in enough detail to suggest whether buying land from family was economically advantageous. Yet Jeffrey Burnham, a white man near forty who was one of New Sarpy's most recent arrivals, suggested that it was a combination of the availability of property and its cost that brought him and his wife to the neighborhood in 2000.

> GO: What made you decide to live here in New Sarpy as opposed to all of the other places you might have lived?
>
> JB: One of the things is cost. Uh, you know, it's, I work for the sheriff's department, so I don't have the greatest of, you know, I don't have that high of an income. So, that was one of the first things. Second thing is, this house actually belonged to my wife's aunt. So, and she passed away, so when she passed away, she knew about this house coming up for sale, and I had been looking for a house, so it just kind of, fell into place.

In Burnham's story, cost, availability, and family connections intertwined, as they did in almost all residents' accounts of how they came to live in New Sarpy. The relative importance of these factors varied among residents, of course. Living near family, something that was a driving factor in Audrey Taylor's and Harriet Isaac's decisions, was seemingly unimportant to Burnham. Economic factors also weighed differently for different residents: Taylor and her husband could almost certainly have bought a house in a nearby town had their home not been available, but others—like Harlon Rushing's teenaged granddaughter and her infant son, who lived across the street from her grandparents in a trailer that the Rushings kept as rental property—might have been in dire straits had they not been able to find a home on land owned by relatives.

Highlighting family ties, acknowledging economic factors, and drawing attention to chance and contingency, stories of how residents ended up moving to New Sarpy after the refinery had already been built undermine the strategic stories told by long-time residents like Irene Masters and Harlon Rushing. Besides showing that the community—at least as it then existed—was not uniformly there first, these kitchen table stories detracted from the strategic stories' narrative of responsible choice. Strategic stories stressed that residents purposefully chose a neighborhood away from industry. Kitchen table stories, on the other hand, showed residents happening into homes in New Sarpy as a result of circumstances only partially within their control. Strategic stories also deleted economic factors from their accounts of residents' choices, implying that they based their decisions to move to New Sarpy on its healthy environment. By reintroducing these factors, kitchen table stories (like Harold Masters's assertion that the plants were where people made their money) opened residents to accusations that money, and not health, was the primary determinant of their choices—and jeopardized the legitimacy of their campaign by showing that many residents had not lived up to the ideal of the responsible chooser.

Understanding Health

New Sarpy's more recent residents scarcely mentioned health concerns in their accounts of how they came to live in the neighborhood. This omission posed an additional threat to strategic stories of responsible choice. Residents who asserted that they moved in before the refinery could claim that they had chosen a healthy community, almost by definition: no refinery meant no refinery pollution and therefore no concern about what that pollution might do to their health. For residents who moved in after the refinery was built, however, the obligations of responsible choice included calculating the health risks posed by refinery emissions.

What a responsible, informed choice to live next to the refinery might entail was described to me by Shell Norco's Health, Safety, and Environment manager Randy Armstrong, a white chemical engineer who used his own decision to live close[9] to the chemical plant at which he worked as an example:

> I live, as the crow flies, less than two miles from the [Shell] facility. . . . I know enough about modeling to know that's about where the peak concentrations are. [laughs] They aren't on the fenceline like everybody thinks. They actually

are slightly downwind, and particularly the elevated releases we have here have a tendency to go about two miles and they mix with the rest of it. So I'm well within the maximum concentration circles. And I know what the risks are. . . . And yet I chose to live here because the school system and the taxes that we paid were good enough that my son could go to public school and I'm a big believer in public schools. And I believe in public schools because that's how people get opportunities to escape poverty and the things that they may not want to do. Therefore, I made a choice to live close by.

For New Sarpy's more recent arrivals to tell a strategic story of their decision to move to a community next door to a refinery—that is, one that established their choice as a responsible one—they would have had to narrate a process of choosing on the basis of scientific information similar to that described by Armstrong. A strategic story might say, for example, that the newcomer investigated the quantities of chemicals emitted by the refinery and the risks that those chemicals potentially posed to their health, deemed them to be acceptable, and *then*, after moving in, found that the refinery had been lying about its emissions or planned to increase them drastically.

Kitchen table stories, however, made no such claims. They were unstrategic in separating accounts of how residents came to live in New Sarpy from accounts of what they knew about the health effects of petrochemical pollution. In the latter, substituting for tales of responsible choice are stories about health and pollution that capture some of the limitations of the information available to residents when they moved to New Sarpy, whether in the 1960s or in the 1990s. In these stories, scientific understanding appears to be a moving target, and experiences of living in the community provide information crucial to residents' assessment of the refinery's effects.

Industrial pollution was not always understood, either by residents or by petrochemical industry scientists and engineers, in the same way it was by the time of CCNS's campaign. Many residents could remember a time when no one considered the possibility that chemicals from the plants might make people sick. As Ida Mitchell explained,

[Y]ears ago, we were really ignorant of the fact that these chemicals cause all these cancer[s]. . . . So we were all ignorant of that fact, and I think most of them [refineries] were ignorant too, 'cause they didn't realize what they were doing, or maybe they didn't care, one of the two.

Mitchell's assertion that chemicals cause cancer would probably have been rejected—or at least highly qualified—by local facility representatives.

Nonetheless, many of them noted the same trend that Mitchell did: they described how awareness of environmental issues had grown over the course of their careers, resulting in significant emissions reductions and more conscientious waste disposal practices. Randy Armstrong, for example, contrasted current practices to those of the early 1970s:

> RA: [W]e have made significant progress since when I started working in environmental, the Cuyahoga River doesn't catch fire any more, and all the fish in Lake Erie are not dead. . . . When I came here, we had direct discharges into the river untreated. Thirty years ago. We were using injection wells for some things. And we were only two miles upstream from the river water intake for St. Charles Parish drinking water.
>
> GO: At that time, was that sort of, would that have been regarded as egregious conduct, or was that sort of best practice for the industry?
>
> RA: That was best practice in 1975. And when I say we've made tremendous progress, those were perfectly legal, permitted activities for both of those kind of incidents.

Residents' accounts also suggest that the changing state of knowledge about the environmental impacts of chemical facilities changed environmental conditions in New Sarpy and surrounding communities. Harold and Irene Masters were among the many residents who told me how terrible it used to smell in Norco. When Harold was a young man, he said, "[By the] time you get close to Norco, you could smell it." Irene added, "Everybody used to say 'we in Norco now, y'all.'"

In comments like these, residents described a general trend of environmental improvement in New Sarpy and the surrounding area. But these improvements did not counteract the growing concern about environmental and health issues. Indeed, a few residents even suggested that other changes at the plants intensified health concerns. Audrey Taylor, for example, blamed the expansion of chemical companies' product lines for increasing the hazards to the community:

> I'm thinking in terms of years ago they weren't making so many things. All right. And so when you think in terms of all the chemicals and things that's being used now, to make the products better, you know, like that, I think you have more dangerous chemicals being used than what you were using back then, because let's face it, they weren't making as many products.

Taylor calls attention to a set of changes—in the complexity and size of petrochemical operations—concurrent with those mentioned by other residents, namely, the growing awareness of the potential dangers of chemicals and changes in environmental practices. These changes combined to produce not only an ever-changing landscape of environmental hazards in and around New Sarpy but also a necessarily shifting landscape of knowledge about them.

Information on which residents might have based their decisions to move to the neighborhood—in particular, information generated by scientific experts—would thus have been a moving target over the course of New Sarpy's development, dramatically different in 1970 than in 1950 or 1990. But the information available also changed significantly once individuals actually moved to the community: as residents, they could draw on their own, and others', experiences of living with petrochemical emissions—information they had no access to as would-be residents.

Recounting their decision to move to the St. Charles Terrace Subdivision in 2000, after Orion was running the refinery, Jeffrey and Debra Trahan, a white couple in their thirties, told me that they moved to New Sarpy with their three school-aged children because the neighborhood was a step up, economically, from where they had been living, two miles away in Destrehan. They had a larger house, more land, and neighbors who were more likely to own than rent their homes. Jeffrey admitted, "I considered a little bit the plant right there, but I didn't think there was any kind of hazard or anything." By the time I spoke to them two years after their move, he told me that he had changed his opinion after noticing that "I wake up sometimes and it's hard for me to breathe, like today, it's a lot harder to breathe than normal."

Other residents not only based their assessments of the health risks of living in New Sarpy on their own sensory experiences; they judged the credibility of scientific information in terms of how well it reflected those experiences. Harriet Isaac, for example, complained to me about industry "propaganda" that claimed that everything was fine, that the plants were not making people sick. When I asked her what would convince her that everything *was* fine, she refused to admit the possibility. To my question, she immediately replied, "I tell you, my nose tells me different." She went on to assert that the data would show the connection and to imply that the only reason it might not do so would be if the right kind of studies were not done: "Let [scientists] come in and live with *us* in the community and not just drive in and out of the community, you know? . . . Come live with us for two or three months. Here." If those making proclamations about health effects were to do so, Isaac insisted, they too would be convinced that the community was being harmed by the chemicals released from the refinery.

While residents like Trahan and Isaac cited their individual experiences as evidence that refinery emissions threatened their health, others drew on community members' collective experience of illness. When I asked Ida Mitchell what made her believe that pollution from the refinery was "poisoning" people, she cited the patterns of disease that she and other residents had observed:

> Because there's too much cancer in this neighborhood. Cause the health study. . . . If you went over the list, how many people had died of cancer within a ten-year period and none of these people being related to each other? I realize genetics play a part in that, but it's got to be something, a common denominator. Too many people. Just in this neighborhood here my mom had leukemia, my sister had three types of cancer, and the man next door had prostate cancer, I guess they took care of that. You know, it's, in every neighborhood, in every street there are people who have died of cancer.

The "health study" to which Mitchell referred was in fact a list of people from the neighborhood presumably affected by pollution, compiled by two residents who spent an afternoon going door to door through the neighborhood, asking (by their own account), "Anybody sick? Anybody died?" On the list, each resident's name was accompanied by a street address, his or her ailment, and, where applicable, the date of the resident's death. Importantly, it was on this relatively informal census of local patterns of illness, and not on formal, scientific research that Mitchell based her conviction that chemicals from the refinery are causing cancer—and her belief that it made no sense to build next to a refinery, knowing what we now know.

Stories like those told by Mitchell and Isaac stress their certainty about chemical health effects: in the face of shifting scientific understandings and industry information, they see their experience as providing ample evidence of the harm done by the refinery to community health. Other residents were less convinced[10] that their experiences trumped experts' claims about pollution and health. While expressing suspicion that their health was suffering from exposure to chemical emissions, their stories drew attention instead to the *un*certainties and areas of ignorance that surrounded the question of how the refinery affected their health. Harold Masters, for example, told me that he was ready to move away from New Sarpy:

HM: Because see, I'm sick already. This ain't gonna help me. . . . If I get away
   from this [refinery] over here, I might live a little longer. Although
   nobody don't know how long they going to live.

GO: Do you think you're sick from the plant?

HM: Well, it could be. It could not be. I don't know for sure. But I'm seventy-
   five years old, and this is the only time I ever been sick like this in my
   whole life. And I been getting around.

In responding to my question, Masters had no studies to cite that might
make sense of his illness. He simply did not know—but imagined that his
health would improve if he moved away from the refinery.

Uncertainty also dominated Audrey Taylor's comments on the potential
of refinery emissions to adversely affect health. Although Taylor had access
to technical and health information presented by experts at meetings of the
St. Charles Community Advisory Panel, of which she was a member, she
still wondered if illnesses in her own family might have been attributable
to industrial pollution. A doctor had told her that the kind of cancer that
killed her daughter was found frequently in people who worked in industrial
facilities or around pesticides, so, she told me, she wondered about that. She
also worried about the health of her two brothers who had worked at Shell
Chemical: the surviving brother had a tumor at the base of his head that
could not even be biopsied without endangering his life. Her everyday experience with chemicals led her to believe that there must be risks associated
with the chemicals produced and released by local industrial plants:

> [W]hen you think in terms of the raw products that are made there [at
> Shell Chemical] to make all of these things that we have, and it's bound to
> not always be so helpful to our health, because when you think in terms
> of even with cleaning products, you got to be careful with inhaling them.

Taylor's understanding of the potential toxicity of chemicals, combined with
the illnesses that her brothers and daughter suffered, left her with, in her
words, "concerns" about the potential health consequences of local industrial
plants. Yet in light of industry officials' reassurances that their emissions met
health-based standards set by the government, Taylor felt that she could only
"wonder" about the connections between pollution and illness, and about
whether regulatory standards were really safe enough.

When New Sarpy residents talked about petrochemical pollution and
the way it affects their health, then, they seldom did so in the language of
responsible choice. That is, they rarely talked, as Randy Armstrong did,

about research on the connections between refinery emissions and health and how that research affected their choices about where to live. Rather, they described how understandings of those connections had changed in the time they lived in New Sarpy—both how scientific and social awareness of the environmental consequences of chemicals have shifted and how their own assessments of the refinery's health effects have been transformed as a result of their experiences of living in the community. Stressing, in some cases, that residents are left wondering about the extent to which their health is affected by refinery emissions, residents' stories also hint at a relative absence of the kind of technical information on which they might base their decisions.

Staying in New Sarpy

Despite the shifting and uncertain nature of knowledge about chemicals' health effects, some of the residents felt that the knowledge that they had now would be sufficient basis for choosing *not* to live in New Sarpy. As Ida Mitchell put it,

> Now that we're all old and everybody's more enlightened in regards to [chemical health effects], I don't know why anybody would want to build next to where they have a refinery. I really don't. I really don't. If I could get a good price for this house, I would leave.

Mitchell's comment hints, again, at a strategic story: residents' decision to live in New Sarpy might have been a responsible choice when they made it. The choice no longer made sense, based on what they knew now, but they were no longer able to make it. With the refinery nearby, no one else would buy their properties, or so the story goes, leaving long-time residents like Mitchell unable to act on new understandings about the risks of chemical exposures.

Kitchen table stories, however, show that residents' reasons for staying in New Sarpy were just as complex as their reasons for moving there in the first place. In fact, Ida Mitchell herself expressed ambivalence about moving away. In the context of saying how convinced she was that chemicals from the refineries and other plants were causing cancer in her neighborhood, she told me that she would leave if she could get a good price for her house. But when I had asked her earlier in our conversation if she had ever considered moving, she had replied,

> I've thought about it. Lots of times. But right now I don't know where I'd go. My son lives out in Saint Amant. But the thing about it, the older

you get, you have to live in the area, even though you may not be what you call close friends, you know, you are friendly with your neighbors. But not chummy. You have to live someplace where you can be sure that maybe if you need somebody, they're there. You know? You can call on them.

Mitchell's belief that the chemicals from the refinery were harmful drove her to consider moving from the neighborhood in which she had grown up and raised her own children. But other factors kept her there: in addition to expecting that she would not be able to get a good price for her home, she worried that other communities could not offer the same kind of social support that she found in New Sarpy. In vacillating about whether or not she would move away, Mitchell raised the same set of issues that shaped residents' decisions to move into the neighborhood in the first place. Despite her health concerns, her relationships with her neighbors and her economic circumstances bound her to the community.

Kitchen table stories about pollution, health, and residential choices also suggested that residents were not, strictly speaking, trapped, as the strategic story would have it. A number of New Sarpy residents told me about siblings and grown children who did leave the community because, at least in residents' telling, they could not abide the pollution. Irene Masters told me that concerns about water quality in St. Charles Parish led her sister-in-law to relocate permanently to the Pacific Northwest. Likewise, after a 1988 explosion at Shell Chemical's Norco plant that damaged his New Sarpy home, one of her sons moved to Atlanta and never returned. Octavia Johnson, an elderly black resident, told me she had a son who was even more directly affected by the plants surrounding New Sarpy: as a seventh grader, he had had such a bad case of asthma that the Johnsons' doctor told them to get him out of St. Charles Parish; he went to high school elsewhere and spent his adult life in other parts of the country.

Factors other than health and safety also spurred community members to relocate. For example, Guy Landry told me that one of his sons, who occupied the house across the street from Landry, would be moving to the nearby community of Montz because his daughter-in-law's parents had a house there that the couple could live in. Indeed, there is every reason to imagine that one-time community members' choices closely resembled those of residents who remained: complex decisions taking family, economics, *and* health into account. Yet the fact that residents did move away, combined with remaining residents' ambivalence about doing so, appeared to contradict strategic stories suggesting that the construction of the refinery had taken away residents' ability to live elsewhere if they deemed the effects of petrochemical plants unacceptable.

Table-Turning Stories

Away from the campaign, in the privacy of their homes, New Sarpy residents narrated how they came to live in New Sarpy—and why they continue to do so—in ways that were dangerous to the strategic stories told during CCNS's campaign. Admitting that residents, in some cases, moved in after the refinery was built, took advantage of economic opportunities, and simply benefited from fortunate circumstances, they undermine the strategic account's emphasis on responsible choices—made by residents when they moved to a peaceful, unpolluted communities and abrogated by the subsequent arrival of the refinery. The kitchen table stories also threaten the moral authority of residents as responsible choosers by showing that careful, scientifically grounded assessments of the potential health effects of refinery pollution did not play an important role in their decisions to move to New Sarpy.

Although dangerous to the strategic story that fueled CCNS's campaign for relocation—that residents made a responsible choice when they moved to New Sarpy and now deserved the right, based on new conditions and understandings, to choose to live elsewhere—kitchen table stories could corroborate a different set of strategic stories, more dangerous to the technical authority of industry experts than to the legitimacy of community campaign. These table-turning stories, told by activists and especially scholars, take aim at the structures that produce environmental injustices and illness from chemical exposures. They also destabilize the notion of responsible choice itself.

Scholarly analyses of environmental injustice stress that communities like New Sarpy end up in close proximity to environmental hazards not primarily because of the choices that individual families make but as a result of persistent social and political structures that systematically produce inequality. They show, for example, how legacies of racial segregation, racist mortgage-lending practices, elite control of local governments, and pro-business government policies act, separately and together, to concentrate environmentally hazardous facilities, from refineries to waste dumps, near communities of color and low-income communities.[11] The structures of inequality that contribute to environmental injustice have been shown to include science and science policy, as well. Scholars have shown how communities' ability to win protective measures on local environmental health issues is severely limited by the combination of regulatory systems that require conclusive scientific proof that pollution is harmful before any action is taken,[12] and mechanisms for selecting, funding, and conducting scientific research areas that ensure that such proof is never really available.[13]

In the context of scholarly analyses of and arguments about the social structures that produce environmental injustice, the economic exigencies and family ties that turn up in kitchen table stories are not evidence that community members failed to act responsibly to choose a healthy environment. Rather, they are evidence that community members' choices are terribly constrained—by poverty, by race, and by the political systems that allocate pollution to the most socially and economically vulnerable areas. A scholar inclined toward structural critique might say of Harriet Isaac's story that she did not *choose* to live next to the refinery, at least not in the sense that Randy Armstrong describes. She was drawn there because she had limited resources and needed the support that her extended family could provide. Why ought she be denied protection from toxic chemicals as a consequence? This kind of strategic story rests not on responsible choice and the moral authority associated with it. Rather, it calls for protection and justice for communities burdened by environmental hazards, questioning the idea that they must somehow be deserving in part by questioning the very possibility of responsible choice in the context of structured inequalities.

As part of analyses of the structures of environmental injustice, studies of scientific knowledge and practice contribute to undermining the logic of responsible choice by unsettling the notion of *informed* choice and, with it, the role allotted to expert knowledge in the responsible choices of nonscientists. Once again, New Sarpy residents' accounts of what was known, what is known, and what is still left unknown about the connections between illness and exposures to petrochemical pollution—that is, their kitchen table stories about health—corroborate these scholarly analyses and critiques, pointing to a series of false premises underlying the model of responsible choice.

Scholars have argued that scientific knowledge is fluid: what we think we know depends on historical context and changes over time.[14] When residents like Ida Mitchell describe the growing awareness of the dangers of chemical exposures around St. Charles Parish's chemical plants, they offer not only an example of science's fluidity but also a window into the concrete consequences of its dynamism. With these stories, they demand to know, How could residents have made informed choices on the basis of information that was constantly changing?

Kitchen table stories also underscore scholarly arguments about the importance of "local knowledge"—that is, knowledge gained through residents' first-hand, place-based experiences—to understanding the effects of refinery emissions.[15] Only after living in New Sarpy for a while do residents notice that their ability to breathe easily varies with air quality, or that an unusual number of neighbors seem to suffer from illnesses that could be caused by chemical exposures. Beyond supporting the argument, popular in the field of science and technology studies (STS), that science ought to respect and incorporate local knowledge rather than dismiss it out of hand,[16] kitchen table stories raise the question, How were residents to make an informed choice about where to locate before having a chance to acquire local knowledge?

Finally, when residents express uncertainty about whether they should attribute their ill health to petrochemical pollution, they position themselves within a knowledge gap, a place where scientific research has been left undone.[17] Knowledge gaps, social scientists have shown, tend to coincide with places occupied by low-income and minority populations[18]—places, that is, like New Sarpy. The absence of scientific knowledge about how chemicals from the refinery affect New Sarpy residents leaves us wondering, To what extent is residents' health endangered by refinery operations? How are residents supposed to choose whether to accept the risks of living in New Sarpy on the basis of studies that do not exist?

Mobilized in the context of research on the structures of science, New Sarpy residents' kitchen table stories can help fuel a powerful strategic story about science and the possibilities for responsible choice. Science is absent (the story goes); it is limited by its inattention to local knowledge; it is necessarily uncertain and shifting. Even if residents had been free to choose any community at all to live in, available scientific data would be insufficient to inform their choices—and local knowledge would have come too late to do so. The story suggests that residents are not irresponsible and undeserving of protection because they moved in after they might have "known" what they were getting into; on the contrary, they could not possibly have known. In questioning the premises of informed choice, this strategic story is most dangerous not to the moral authority of residents but to the claims of scientific experts, and their status as people who can inform residents' choices.

Enterprising Stories

A month after CCNS's settlement with the refinery, Orion community relations manager—and black New Sarpy native—Henry LeBoyd offered his story of residents' campaign and their continued presence in the community. Throughout the campaign, Orion officials had maintained that the refinery's operations had no adverse effect on air quality and certainly did not harm New Sarpy residents' health. According to LeBoyd, who repeated the claim to me between bites of Chinese food at a Destrehan restaurant, the fact that CCNS members had been living in New Sarpy proved it—proved, in fact, that even residents themselves did not believe that Orion was a threat to their health. Baffled by his logic, I asked him to elaborate. Look, LeBoyd explained, if you truly believed that your health was in danger you would not stick around. You would get yourself and your kids out of that environment to somewhere that you felt safe. But what about residents' argument that they could not go because they could not get a good price for their houses? I asked. That would not matter, he insisted. If you were really worried about your health, the money would not matter; you would find a way.

LeBoyd's story was an enterprising one. Like strategic stories, it hinged on the idea of responsible choice: residents decided where they would live, he suggested, on the basis of their assessment of the risks. But it incorporated a heightened version of responsible choice, one characteristic of "enterprising individuals"—the model citizens of contemporary, neoliberal democracies who actively seek work, products, and ways of life through which they can better themselves.[19] In particular, LeBoyd's story stressed the autonomy of the enterprising person—the idea that, although their proactive choices are guided by expert advice, enterprising individuals make their choices unhindered by obstacles that cannot ultimately be surmounted through personal initiative[20]—by suggesting that New Sarpy residents who truly feared for the health of their families would somehow manage to move away. Attributing enterprising individual status to residents leads LeBoyd not to question the responsibility of their choices but to accuse them of lying about their motives. CCNS members have made a calculated choice to stay in New Sarpy, LeBoyd suggests, and through their campaign are actively pursuing their economic interests with health as a smokescreen. In keeping with the neoliberal presumption of personal autonomy, his enterprising story disallows the possibility of structural constraint. Residents simply could not be trapped in New Sarpy; if they were there, it was because they chose to be.

The stories told by other industry representatives—stories about residential choices and the health consequences of petrochemical emissions—were similarly enterprising. They, too, ignored the possibility of structural constraints on residents' (and their own) choices, especially the systemic limitations of scientific knowledge that called into question the possibility of *informed* choice. Instead, industry engineers' and scientists' stories drew on their own, taken-for-granted status as enterprising individuals to bolster their scientific claims—and to sidestep the dangers that residents' kitchen table stories, strategically deployed, posed to their technical authority. In addition, by presuming *residents'* drive for personal fulfillment through acts of choice, industry experts' enterprising stories defined a role for their knowledge in those choices—a role that further obviated critique by allowing expert knowledge to be essential and uncertain at the same time.

Validating Expert Knowledge

The engineers and scientists who managed petrochemical plants in New Sarpy and Norco, people like Randy Armstrong and Orion's Jason Carter, were defensive about allegations that their pollution made people sick. They were particularly rankled by allegations by angry residents that they did not live near the facilities in which they worked because they themselves felt that it was not safe to do so.[21] While industry experts referred to various forms of scientific data in answering these charges, the primary evidence that they offered to contradict activists' assertions that the plants were not safe were their own choices to work in the plant or even live nearby.[22]

Mitchell Mobley,* a white plant manager for a petrochemical facility in the region, was unconvinced by activists' assertions that plants like his were a health hazard. His company had done health studies of its employees, he said, that showed that they lived long, healthy lives. While he had not seen any studies of community health, he was impatient with activists' reasoning that, in his words, "you don't live here, so you must think it's bad." His willingness to work at the facility, he suggested, provided ample evidence that it was safe.

> It isn't that we think these are bad places, if I did, I wouldn't work in it. I'd be stupid to. You think we have carbon filters here that are . . . ? No. We're breathing the air that everybody else breathes. Do we get oil on our car if we have a relief valve pop to the atmosphere? Yeah. Do I like it? No. Would that keep me from living here if there was a community that my wife and kids have the things they're used to? Of course I'd live here. It's silly to think otherwise.

Although Mobley's confidence that it was safe to be around the plant was arguably based on (limited) scientific data, he mobilizes primarily his own choices in answering community allegations that his plant is hazardous. He offers his presence at the plant, his willingness to breathe the air and take the consequences of accidents, even his history of having lived near other plants at which he was a manager as evidence that they are not "bad places."

The idea that Mobley's own decisions could be a kind of evidence, supplementing scientific data about the health risks associated with petrochemical plants, relies on the assumption that Mobley is an enterprising individual, determined and able to make informed choices. Jason Carter, a high-level engineer at Orion, asserted his status as an enterprising individual even more explicitly in telling me that his plant did not have environmental problems. While Orion did not, at the time, conduct any ambient air monitoring, Carter referred to the monitoring program being conducted by his counterparts in Norco, which was showing that air quality near the refinery there was at least as good as the state's ambient air standards. But, saying that these results did not surprise him, Carter offered his own choices as even more compelling evidence that his plant was not hazardous to the surrounding community.

> I'm not a stupid man, okay? I've got three small kids and a wife. I'm certainly not going to live or work anywhere where my health is at risk. I'm sorry, I'm a little too selfish for that. Okay? And I wouldn't be in this plant if I felt I was being exposed to benzene or SO2, I wouldn't do it. For my own self-preservation. But I know it's not an issue. I know this plant is clean, I know there's not an emission problem here, I know there's not an exposure problem here.

In making his personal decisions part of a case for the safety of the refinery, Carter positioned himself as precisely the kind of person who makes informed choices in pursuit of his (and his family's) well-being: he is "selfish" and interested in "self-preservation." Further, because he is "not stupid"—that is, because he is capable of understanding risks and taking them into account when making decisions—one can look at his choice to work at the refinery and feel reassured that the plant really is clean and safe.

Focused on industry engineers' personal choices, these enterprising stories obscured the fact that the data that underpinned their choices were scant and easily contested. The health studies that Mobley cited focus only on workers, are conducted by industry, and are not made publicly available. Carter referred to air monitoring being done near another refinery without acknowledging their significant differences, including his refinery's history of having—by his own admission—an environmental record regarded as egregious throughout the industry. And in telling me that he made the choice to live just two miles from Shell even though it put him "well within the maximum concentration circles," Randy Armstrong invoked not measurements of chemical concentrations but computer models of the way chemicals disperse in the air—models whose quality would depend significantly on the accuracy of assumptions, also not public, made about emissions rates and sources of emissions, among others.

But these fragile data remain in the shadows of experts' stories about health and safety at chemical plants. Their choices to work at and live near the plants are in the spotlight, upstaging reference to the information on which those choices were made. Drawing on the presumption that they are enterprising individuals, industry engineers are able to turn attention away from data that would be easy to contest. They can ground their arguments instead in something much harder to quarrel with: their dedication to looking out for themselves and their families. In this context, information ceases to appear as a justification for choice; rather, through the presumption of enterprising personhood, choices become a way to shield contestable information from challenges.

Validating Residents' Choices

Industry scientists and engineers told enterprising stories in interviews with me as part of imagined arguments with environmental justice activists. They told enterprising stories when they spoke directly to community members, as well—although in that context their stories about chemical emissions and their relationship to health drew on not their own but residents' status as enterprising individuals to establish experts' authority. As enterprising individuals, residents were presumed to rely on the information provided by scientific studies. Yet, with the studies cast as inputs to choice, their uncertainties or limitations became matters for individual judgment—and not grounds on which experts' claims could be criticized.

One study in particular was presented perennially to residents of New Sarpy and surrounding communities in the hopes of allaying their concerns about chemical plants' effects on their health. The Louisiana Tumor Registry's reports on "Cancer Incidence and Mortality in Louisiana," first published in 1983 and updated regularly since, are among the studies frequently cited by representatives of petrochemical facilities claiming that their operations do not harm nearby communities. As a result, they arranged periodically for staff members from the Louisiana Tumor Registry (LTR) to present their findings to community representatives at meetings of local Community Advisory Panels (CAPs).[23] Minutes from the meetings of the St. Charles CAP, which included residents of New Sarpy and Norco, report on presentations of the LTR's study on several occasions over the course of the CAP's thirteen-year history (1992–2005), including at the November 20, 1997, and September 28, 2000, meetings.

The chief finding of the study, emphasized by industry representatives and reflected in the presentations by the LTR's Dr. Vivian Chen, was that the incidence rate of cancer in Louisiana's "Industrial Corridor" is, for most demographic groups and forms of cancer, no higher than the national average. Minutes from the 1997 meeting, for example, report that "[i]n the Incidence Study it was found that, except for Lung Cancer, the rates of cancer in the River Parishes[24] were similar or lower than the rest of the country." The information was clearly meant to reassure the audience about their choice of residence. According to the minutes, Chen's presentation of statistics on incidence rates at the 1997 meeting opened with the question, "Is the term 'Cancer Alley' fact or fiction?"—referring to the term that environmental activists coined to highlight the (according to them) rampant health problems in the industrial corridor. By debunking the popular myth that cancer was more prevalent there than elsewhere in the nation, the presentation suggested that residents need not worry that they were increasing their risks of disease by living near a petrochemical plant.

But in the context of Chen's presentations, the finding that cancer rates were not elevated in the industrial corridor had a second implication for residents: personal lifestyle choices were extremely significant for their health. Chen began her presentations by noting how prevalent cancer was. After comparing the cancer rates in various Louisiana regions to national statistics, Chen returned to the most common forms of cancer, discussing their prevalence, prognoses, and means of prevention. In talking about lung cancer in her 2000 presentation, for example, meeting minutes report that

> Dr. Chen used the example of lung cancer deaths in Louisiana to [lives lost in] a jumbo jet crash.[25] The FAA spends millions of dollars investigating a crash, and in many cases it can be attributed to human error. However, in the case of lung cancer, there is no human error involved—it is human choice!

The choice to which Chen referred was primarily the choice to smoke tobacco: she continued by outlining the increased risks of lung cancer faced by smokers. Acknowledging the magnitude of the problem and attributing it to "human choice," Chen suggested that her listeners were right to be concerned about their risks of cancer, but that they should focus on the actions that they themselves could take to reduce their risks—for example, not smoking.

Chen geared the discussions of other cancers, as well, to informing residents' behavior. Her 2000 presentation acknowledged that little was known about what causes or how to prevent several common cancers, including prostate, breast, and colorectal cancer; as a result, she stressed, regular screenings were important to allow for early diagnosis. Her conclusions suggested that tobacco use and diet accounted for two-thirds of cancers in Louisiana (the other one-third was allotted to "environment and other factors," presumably including genetic factors), and she ended with a list of recommendations for how residents should behave to reduce their risks. They could "work together for a clean environment," but seeming to acknowledge the enormity of that project ("do not be pessimistic"), she urged residents to realize that "we have some control individually" and thus take steps to manage their own risks: "do not smoke at all"; "drink moderately, or not at all"; "eat healthy"; and "follow guidelines of cancer screening."

Chen's presentations of the Louisiana Cancer Incidence study transformed a discussion of health in the communities around one of the nation's largest concentrations of petrochemical facilities into a recitation of standard medical advice about how to lower one's personal risk of cancer. In doing so, it reinforced residents' status as enterprising individuals. Their choice to live in a heavily industrialized area, the study reassured residents, was not an irresponsible one: environmental risks to their health were minimal. On the other hand, Chen suggested, enterprising residents would want to take action on the "real" risk factors, which were assumed to lie solidly within their control.[26] As an aid to responsible action, the Tumor Registry's study became relatively difficult to quarrel with: an enterprising individual would not eschew data that pointed the way to better lifestyle choices in order to insist that health problems in the region were caused by environmental factors outside her control.

Industry representatives' presentations of a second study similarly cast residents as enterprising individuals, positioning experts' information as an important input to choices that residents ultimately had to make for themselves. In the spring of 2003, Shell and Motiva in Norco held a series of meetings to share the results of the first phase of their "Air Monitoring...Norco" (AMN) program. Designed to answer Norco residents' questions about air quality in the town, which was sandwiched between a Shell Chemical plant and a Motiva refinery, the program collected data about levels of toxic chemicals in the ambient air at six different sites; the first phase represented the first four months of sampling, conducted in the last part of 2002. Three key findings were presented at two special community meetings in Norco, at a meeting of the Norco-New Sarpy Community-Industry Panel, at a media briefing, and in a flyer distributed to all Norco households. The study concluded that the measured levels of chemicals were within the state's ambient air standards; that air quality was fairly uniform across Norco; and that, as reported in the flyer, "Norco's air quality is similar to that of other cities."

With the third claim in particular, the engineers and scientists involved in AMN invoked the enterprising individual as a way of giving their technical information meaning. According to one engineer, the team compared measurements from Norco with those taken in other cities because they had wanted to let people see what the air was like in Norco versus what it was like in Santa Monica—presumably another place where Norco residents could have chosen to live. Randy Armstrong made the connection even more explicitly in the media briefing. Shell could not say, he admitted, that you would not be able to find cleaner air than in Norco. But that was likely to be somewhere out in the country. It would be up to individuals, he went on, to look at the data that Shell had gathered and make the decision about whether the amenities of an urban area like Norco outweighed its higher—but still within regulatory standards—levels of air toxics.

Armstrong's explanation of AMN's findings drew on the presumption of enterprising personhood in a way that, once again, drew fire from experts' scientific data. The finding that Norco's air quality was similar to that of other "cities" was arguably spurious: with thirty-seven hundred people and three traffic lights on two and a half square miles, neither Norco's population nor its population density remotely compared to that of Houston, Burbank, Minneapolis, or any of the other cities whose air quality data engineers used in arriving at their conclusions. Yet by suggesting that, as enterprising individuals, residents would look at the data and make their own choices, experts like Armstrong suggested that the specific comparisons that he and his colleagues made were only partially relevant. What residents really had to decide was whether they wanted to live somewhere "out in the country" (like, perhaps, Minneapolis, where air quality was much better than in Norco), or whether they could better pursue their well-being by living in Norco, despite its higher (but still safe, at least when compared to the regulatory standards) chemical concentrations.

In enterprising stories, then, the presumption of responsible, autonomous personhood becomes a bulwark against challenges to industry scientists' and engineers' technical authority. Scant as it is, petrochemical facilities' data on the health effects of their operations might be subject to criticism. But facility managers' decisions to expose themselves and their families to the same pollution that residents experience become weighty evidence that the plants pose no threat to health when it is taken for granted that managers are enterprising individuals. And when it is presumed that residents can and do pursue their well-being by making responsible choices, the partiality or uncertainty of scientific studies becomes less important than their ability to provide information on which enterprising individuals can act, information that advises them that they can ensure their health by not smoking or, if they are exceedingly risk averse, by moving to Minneapolis.

Choosing a Story

"Who was there first?" was inevitably the first question I would be asked by students when I introduced them to New Sarpy and Norco as part of an extended case study in my engineering ethics classes at the University of Virginia. *My* story was that both industry and residents were always already there. The New Orleans Refining Company and Shell Chemical and the Good Hope Refinery each built their facilities in an area where there were already industrial facilities *and* residential communities. Residents moved into an area with some number of industrial facilities, but had seen them expand and proliferate during their tenure in the community. Communities and industrial facilities were so close together now, I would tell my students, because of a gradual process of encroachment on both sides of the fenceline.

"But what difference does it make?" I would ask them. The idea that people should not have chosen to live there if they knew there was a chemical plant that could harm their health—an idea that seemed terribly logical to these mostly well-off, mostly white, 21-year-olds—neglected a couple of important facts, facts that I learned over New Sarpy's kitchen tables as a well-off, white 27-year-old. People do not necessarily choose where they live in a calculated way; they decide to live close to their families, for example, or to buy wherever they can find a house for sale that they can afford and that will hold all their stuff. Maybe more importantly, I would say to them, thinking of their future careers as engineers, the kind of information that you would want to have to figure out whether you might get sick by living someplace just did not exist. It was incomplete, uncertain, contested—and, where it existed at all, probably hard to get to and even harder to understand. Besides, I would conclude, why should it be that, just because people *chose* to live someplace, it was okay for them to get sick from something put into the environment by big corporations?

The story that I told my fledgling engineers—that industry and residents were both always already there; that choices are complex and constrained rather than calculated; and that it makes no sense to think of existing science, or perhaps science at all, as complete or certain enough to adequately inform the "choice" to live next to a hazardous facility—is as accurate as any. Rooted primarily in residents' kitchen table stories, it is also clearly dangerous—but it is not clear for whom it holds the greatest threat. It brings to light structural constraints on residents' choices, especially the near impossibility of responsible choice informed by scientific knowledge. In that sense, it is dangerous to the environmentally unjust status quo in general and to the authority of the scientists and engineers who help maintain it in particular.

My story, however, is also dangerous to the moral authority of residents as good neoliberal subjects. Unlike the strategic story told during CCNS's campaign, it does not hinge on the idea of responsible choice. In fact, by emphasizing the social inequalities and systematic limitations of science that make informed choice an idealization at best and a dangerous fallacy at worst, my story denies residents their status as responsible choosers. With respect to the strategic story, my denial of responsible choice arguably only intensifies a preexisting ambivalence: in the strategic story, residents are hypothetically responsible choosers, but are prevented from making the choices that they wish to make. In the context of industry officials' enterprising stories, however, my story is terribly dangerous to residents. Because these stories deny the possibility of systemic obstacles to responsible choice, my suggestion that responsible choice, at least as popularly understood, is not possible for residents effectively strips them of their status as "enterprising individuals" and the moral authority that comes with it.

In our neoliberal moment, stories that afford people the status of enterprising individuals, of autonomous persons, of model citizens are compelling. Being able to freely choose their own circumstances figured powerfully into New Sarpy residents' conceptions of themselves, myriad structural forces notwithstanding. After telling me how she had built a home on her grandmother's land so that her children might be raised around family, for example, New Sarpy native Harriet Isaac told me,

> I really believe that this plant is vital to the national energy. . . . Do I think it's more important than my health? No. But, you know, at this point in time . . . I chose to build my house here and live here, you know. So, you know, one day I'll choose to move.

By neglecting the variety of contingent factors that brought her back to New Sarpy in the mid-1980s, Isaac reasserted her control over a choice that involved, as she framed it, a balance between personal risk and public good. Through this account—and not, importantly, the strategic story in which residents like Harlon Rushing claimed that they had become unable to choose to move—Isaac was able to claim status as a fully autonomous, responsible individual.[27]

It is in part because of this self-conception, because of the generalized obligation to understand oneself and one's actions in terms of a heightened, neoliberal notion of responsible choice, that the ideal of the enterprising individual becomes so powerful in industry scientists' and engineers' work to reestablish their expert authority in the face of grassroots challenges to shaky science. Their enterprising stories do not quell residents' suspicions that they are being harmed by petrochemical plant emissions. They do not permanently silence opposition. Nor do they completely neutralize critiques launched by scholars and activists, which call attention to the limitations of the evidence on which the petrochemical industry justifies its continued operations. But they do provide a basis for sidestepping some of those critiques in a way that allows for the reconstruction of industry officials' technical authority: where scientific data is scarce and contested, industry officials, as enterprising individuals, can offer their own choices as a kind of evidence that is difficult to challenge. Moreover, these stories of experts' own responsibility are even more resistant to critique because they extend the presumption of enterprising personhood to residents, as well. By offering residents moral status that strategic stories subtly deny, enterprising stories significantly raise the costs—for residents if not for their allies—of challenging experts' authority. A looming figure in the neoliberal landscape, the enterprising individual thus serves as a resource for petrochemical industry scientists and engineers seeking to turn back challenges to their scientific claims and construct themselves as authorities on the environmental and health impacts of their facilities.

3

Noisome Neighbors

> Fordylson looked at me hard and didn't have to say what he was
> thinking. He glanced down at the ground between his smooth-toe
> lace-ups. "And clean up your yard."
> "What's that got to do with anything?"
> "It's got everything to do with everything."
> —Tim Gautreaux, "Welding with Children," 1997

> Now Gwen you ought to see [Dane's] house. It is gorgeous. He had
> cement poured in most of his yard. On both sides of his garage
> with 2 patio covers on each side [and] a patio cover over his side
> porch. Then his vinyl fence is so pretty. In the back part of his yard
> he has 6 ft. all white.
> —Myrtle Berteau, New Sarpy resident,
> letter dated September 10, 2003

Perhaps you would like to see for yourself the sites of these battles over industry's obligation to its neighbors, over the sustainability of petrochemicals, over the dominance of expert knowledge? Find I-310, a spur off the cross-country interstate I-10 about fifteen miles outside of New Orleans, and make your way south to the River Road (LA-48). Turn right off the exit ramp, and a grassy slope several stories high follows the road on your left—that's the levee, blocking your view of the Mississippi River just beyond. On your right, pass the streets that comprise Destrehan, the largest town in St. Charles Parish, and myriad parish institutions: a recreational area with ball fields, a branch library, the junior high and high schools, the Ormond plantation house (now home to a weekly farmer's market), the Catholic church from which the parish takes its name, a police station, and a post office.[1]

Nothing will mark your transition into New Sarpy—by the time you see the do-it-yourself carwash, you will know you are there—but two-thirds of the way through the town, you may notice a flowerbed planted into the slope of the levee. A sign in it announces that you have reached the four-street-by-two-block neighborhood that campaigned against Orion in 2001 and 2002: "St. Charles Terrace Subdivision," it reads, continuing in smaller print below, "Sponsored by Valero Refinery."

Still you might not see what all the fuss was about. But turn down St. Charles Street and roll down your window. Maybe you will notice a slight rotten-egg smell or a waft of motor oil. By the time you have gotten to the "back" block beyond Short Street, though, you can look to your left across vacant lots and see it: the refinery fence, the storage tanks that would each cover a football field, the tall stack topped by a flare—as well as additional processing units built since Valero took over the refinery in June of 2003. As you ride back out to "the front,"[2] notice the homes: a couple are crumbling, tar-papered over, with trash in the yard. Many are weathered and aging, but tidy and well kept. And the gleaming one with four steps up to it—look closer; that is actually a double-wide trailer, all fixed up. When you cross Short Street again, look for indications that you are going from an all-black part of the community to a nearly all-white part.[3] You might see the change, if you happen to spot people out on their porches or in their yards, but there will be no trace on the houses themselves.

But keep riding up the River Road and you will really start to see. After a row or two of storage tanks, towering industrial units, connected by high-flying pipes, come thicker and faster. Bundles of pipes and, in one place, a rectangular silver conduit reach over the road to the transport ships that dock on the river's bank. If you are not too bedazzled by the monstrous technology, you may note at the edge of the road a couple of early-twentieth-century buildings—formerly a school and a church—now bearing "Valero" signs. Between them, Prospect Street snakes through the refinery out to Airline Highway, the pre-interstate route back to New Orleans; Prospect's S shape is a result of a mid-1990s battle over where the refinery would locate a new coke conveyor unit.

After a mile or two, Valero signs give way to Motiva signs and then to Shell signs as the processing units and overhead pipe racks continue. And then you hit a stop light, the first since the entrance to the upscale Ormond subdivision—where Randy Armstrong and other top industry officials live—back in Destrehan. Turning at the light takes you down Apple Street through the town of Norco. Its sights are similar to New Sarpy's, though on a grander scale.[4] Apple Street, which will also take you all the way out to "the Airline," crossing two sets of railroad tracks along the way, is lined with locally owned businesses (all of New Sarpy's commercial property, in contrast, lies along the River Road). The two streets between Apple and the Shell/Motiva fenceline[5] comprise many vacant lots, warehouses and other small industrial properties, and relatively run-down homes; turn left off Apple Street, however, and you will find yourself amid blocks and blocks of tidy, moderately sized ranchers.

Driving down First Street will take you through the undeveloped Gaspard-Mule tract into Diamond, the historically African American part of Norco that won relocation from Shell Chemical in 2002. Diamond's four streets, especially Washington and Cathy, the two closest to Shell's west site, are now mostly just green space sprinkled with the occasional house or trailer, ranging from ramshackle to meticulous, belonging to residents who chose not to move. A commemorative sign at the corner of Washington Street and River Road, at the edge of a block that is nothing but grass and trees, reads "Diamond Community—Established Early 19th Century." Across Washington Street, Shell Chemical's processing units crowd the fenceline and continue to emit the gases that Concerned Citizens of Norco charged with threatening their health.

* * *

"Well, what is your opinion about the community . . . do we have a good community? . . . as an outsider, what is your view?" The question was put to me by white Norco resident Milton Cambre. I stammered through an answer (and what would you say, now that you have seen the place?), initially surprised that my opinion would matter to Cambre, who had told me earlier in our May 2003 interview that he had "always thought Norco was a nice town." But in fact Cambre's concern about an outsider's view represented an important dynamic that helped to shape the outcome of CCNS's campaign (and that fueled bitterness among whites in Norco about Concerned Citizens of Norco's campaign): the need to maintain the town's public face as a "good community," or a nice place to live.

Despite their ultimate goal of relocation, CCNS members asserted throughout the summer of 2002 that they were working to make their neighborhood a nice—or, perhaps, a nicer—place to live: through their Clean Air Act lawsuit, they sought improved environmental quality that would benefit everyone, especially those not interested in moving even if a relocation program were offered. They accused Orion of making their good community, in the words of Harold Masters, "bad like it is now," and—in the wake of Orion's July 2002 Community Improvement Program offer—dividing their "tight-knit" neighborhood. But as the summer wore on, a rival group was established that charged CCNS with "tearing down" the community. The new St. Charles Terrace Neighborhood Association (SCTNA), formed with support from Orion, refused the notion that New Sarpy was in any way "bad" as it was. The group articulated a program for "building up the community" that ignored environmental issues and instead actively sought investment from local industry.

However self-interested their motives (CCNS members accused SCTNA's founders of becoming involved in local issues only because they wanted to get the cash payments being offered to families in Orion's Community Improvement Program), SCTNA's defense of New Sarpy and agenda for improvement exemplified a distinctly neoliberal approach to community development in which outsiders' perceptions matter a great deal—an approach that geographer David Harvey terms "entrepreneurialism." Similar to other neoliberal forms, entrepreneurialism moves away from direct government investment in communities and instead seeks to promote urban development through partnerships with private corporations.[6] In evidence in the boast, "Sponsored by Valero," on New Sarpy's eventual roadside sign, successful entrepreneurialism rests in part on outsiders' perceptions of communities as "good" places: the focus on public-private partnerships, Harvey and others point out, leads to competition among municipalities for corporate investment and consumer dollars,[7] competition in which a community's image can be decisive.[8]

The logic of entrepreneurialism is also implicated in the construction of scientific knowledge and expertise.[9] Fenceline communities like New Sarpy depend on their good image to attract new businesses and ensure the continued saleability of homes;[10] moreover, their most important private-sector partners are the large industrial facilities next door. The combination provides significant incentives for residents to accept petrochemical industry expertise. Resident-generated evidence of environmental degradation and ill health—bucket results showing a pattern of high levels of air toxins, for example—can tarnish a community's image and make it unappealing to potential home buyers and other investors;[11] in contrast, claims by petrochemical industry scientists and engineers that environmental and health effects of plant operations are minimal, although judged spurious and self-serving by activists, can bolster or even recuperate the image of a community dependent on others' good opinion for continued development. Many formal mechanisms for industry investment in fenceline communities, furthermore, incorporate the assumption that the scientists and engineers running chemical plants do so in a manner that ensures community health and safety—an assumption that community members must tacitly accept if they are to benefit from industry's largesse. The logic of entrepreneurialism, then, encourages acceptance of expert ways of knowing that fail to acknowledge the systemic threats posed by industry, no matter how problematic. As part of the neoliberal terrain on which campaigns like that in New Sarpy are conducted, it thus offers a resource for experts eager to discourage and discredit citizen science efforts that challenge their authority on environmental and health issues.

Entrepreneurialism in St. Charles Parish

An unincorporated town with a population of seventeen hundred people in a couple of square miles, New Sarpy is not the first place that springs to mind as an example of urban entrepreneurialism. Yet entrepreneurialism is clearly at play in St. Charles Parish more generally. In particular, the St. Charles Parish Economic Development Commission makes it its mission to attract, grow, and keep businesses in the parish. The commission's pitch to companies highlights some of the strategies used by local governments in their bids to attract investment—and hints at the importance of image, or reputation, in marketing an area to prospective investors. *St. Charles Parish: The Best of All Worlds*,[12] an online video that echoes the pitch outlined to me by a commission staff member in 2003, advertises easy access to all major modes of transportation and shipping, thousands of acres of land available for development, and a highly skilled labor force, evidencing one of the primary strategies for entrepreneurialism identified by David Harvey, one focused on "the creation [or] exploitation of particular advantages for the production of goods or services."[13]

In representing the parish as well suited to industrial production, the video implicitly tackles well-worn stereotypes. For example, it gives the impression of great diversity in local business against the (perceived and actual) dominance of petrochemical companies in the parish and the region: of the five business spokesmen[14] featured in the video, only one represented a petrochemical facility, and examples of success mentioned chemical and energy companies as just two among a list that featured "logistics" companies (e.g., FedEx), "maritime leaders," a beverage distributor, a men's accessory company, and a food technology incubator. The video also makes a point of trumpeting "ethics reform" in Louisiana's state government and cites a number one ranking by the Center for Public Integrity for legislative financial disclosures as evidence that Louisiana "government works effectively"—an important selling point, presumably, in the context of the state's long-standing reputation for political corruption.

But the video's pitch is not limited to business considerations, at least not as normally understood. It also emphasizes living conditions in the parish—pointing to an aspect of entrepreneurialism whose relevance to industrial investment is not necessarily obvious. The video, in fact, opens with reference to quality-of-life concerns: over an image of a golden sunset, the narration begins,

> Big business doesn't only happen in big cities. What if you could take just the best parts of your job and relocate to a place where recreation is valued as much as work. Where you can see the stars at night. Where your children attend the best schools in secure neighborhoods.

The end of the video returns to these themes, touting the possibilities for outdoor recreation year-round and the ability for children to get a "superior education without burdening families with the cost of private schools." The emphasis on these themes suggests that the ability to attract investment to the parish—at least in the view of the St. Charles Parish Economic Development Commission—depends on being able to represent the area as a place that well-educated, affluent workers who move around for their jobs would find appealing: a place where they can bike, golf, and camp; a place where they can enjoy good food and cultural events; a place where they feel safe and able to provide for their children's education. Thus even in a place like St. Charles Parish, where industrial development (as opposed to, for example, tourism or consumer spending) continues to be seen as the engine of economic growth, entrepreneurialism entails not only demonstrating a good business climate but also presenting an image as a nice place to live.[15]

Specific St. Charles Parish communities like New Sarpy and Norco were not themselves engaged in wooing business investment—since they did not have their own governments, that task belonged to the parish. However, companies like Valero and Shell wanted to be able to attract specialized, high-level workers to their facilities, and residents' home values depended on outsiders' interest in relocating to the community. As a result, an entrepreneurialism compatible with that of the parish government was manifest in local projects for community improvement. Undertaken both by civic groups and by industrial facilities, but often involving some element of collaboration between the two, such projects aimed to improve the quality of life for residents. Yet these quality-of-life improvements, in many cases, simultaneously worked at a symbolic level as well, reaffirming the meanings that residents attached to their communities and enhancing the image they presented to outsiders.

Improving Norco

Community-improvement activities were especially lively in Norco, where both residents and industrial facilities initiated projects to better the town.

Residents' efforts were concentrated in (though by no means limited to) the Norco Civic Association (NCA). Formed in 1997 and comprising several dozen active residents, most of them white, NCA described itself as "an organization that will focus on and promote issues that will bring about improvements in our community . . . an organization that will pursue and advance the objectives we believe are necessary to improve the quality of life for the people of Norco." Its projects ranged from persuading the parish to improve traffic signage and lobbying for changes to zoning laws to limit trailer homes in the community, to landscaping public areas and holding an annual community Easter egg hunt.

As the two largest industrial facilities in the community,[16] Shell Chemical Norco and Motiva Refinery were frequently solicited for contributions for community projects and often agreed to help out. But the companies also had their own community-improvement programs, which were organized under the umbrella of Good Neighbor Initiative (GNI), launched in 2000. Honored in 2003 by Shell Chemicals as an example of best practice across the global corporation,[17] GNI was divided into environmental, community-health-and-safety, and quality-of-life project areas. While the environmental and health and safety areas focused largely on technical issues of plant performance that affected Norco residents—reducing emissions, flaring, and noise, for example, and improving emergency response plans—the quality-of-life area engaged with community needs and desires directly. Under its auspices, Shell and Motiva sponsored "educational initiatives," including upgrades to the Norco Adult Learning Center's facilities and programs; developed a greenbelt area between the plants' fencelines and the nearest residences; and established the Norco Community Fund, an endowment to support community- and nonprofit-led projects to improve the quality of life in Norco.

The projects funded by the Norco Community Fund in its first year suggest how community image was tied up in efforts to improve quality of life for residents. Awarded in a competitive process, five of the nine grants went to infrastructure improvements at schools and nonprofits: plumbing renovations at the Norco Adult Day Care were funded, for example, as was the automation of the Sacred Heart Schools' library system. The St. Charles Parish public schools were also awarded money to run an elementary school science camp. While these six projects could be argued to bear directly on the quality of services available to residents through various local organizations, the remaining three awards were made to projects whose impact on quality of life is less immediately apparent: NCA received a grant for the annual Norco Christmas parade, which the organization had revived in 2001 after a hiatus of many years; the River Road Historical Society won money to hold a Norco Community Heritage Day; and the St. Charles Historical Foundation was funded to collect oral histories focused on two now-closed Norco schools.

These three projects all worked on Norco's image, both in the minds of residents and in the eyes of the outside world. The Christmas parade and the Norco Community History Day were both fun, free events for residents— and the parade in particular was rare in being equally accessible to both blacks and whites.[18] Like the other kinds of festivals put on by entrepreneurial cities,[19] both events also had the potential to attract visitors from outside Norco, and, indeed, residents of neighboring towns in St. Charles and St. John parishes attended the Christmas parade. Moreover, reaching across the community and beyond, these public spectacles marked Norco as a town of note, a community large and wealthy enough to put on a good time for its residents and neighbors.

If the Christmas parade helped make a name for Norco as a fun, vibrant place, the Community History Day and the oral history project funded by the Norco Community Fund asserted Norco's significance as a historical place. Many—perhaps most—Norco residents already took pride in their town's history: the River Road Historical Society's museum, located on Shell Chemical's east site, celebrated the codevelopment of the Shell facilities and the town of Norco; many Diamond residents, for their part, claimed among their ancestors participants in the largest slave revolt in U.S. history.[20] But Norco is also located in an area that tourists visit for its historical sites: the restored Destrehan plantation house, five miles downriver from Norco, is probably St. Charles Parish's leading tourist attraction, though guidebooks like Mary Sternberg's *Along the River Road* find sites of historical interest throughout the parish.[21] Projects like the Community History Day and the oral history project, then, built up the community not only by appealing to residents' pride and interest in their history but also by bolstering the community's claim to be regarded as one of the region's historical sites.

But it was not only community-improvement initiatives that were animated by a particular kind of small-scale entrepreneurialism. Shell Chemical's purchase of residential properties in Norco was also represented, and justified, in terms of the intertwined issues of quality of life and community image. In September 2000, after years of activism by disgruntled residents of Diamond, Shell announced its Voluntary Property Purchase Program (VPPP). The program offered to buy the homes on the two streets closest to Shell Chemical in Diamond, as well as on the two streets closest to Motiva and Shell's east site in the white part of Norco. Residents were not included in developing the initiative, and CCN and its allies criticized the program as an attempt to silence the most vocal activists in Diamond, most of whom lived on the streets included in the program.[22] But representatives of Shell Chemical explained their reasoning in terms of quality of life in the community. The company wanted to establish a "buffer zone" between their facilities and the community, GNI manager David Brignac told Steve Lerner in 2002. Why not include all of Diamond in the buffer zone? Extending the program to all four streets, Brignac explained, would require including two more streets on the other side of town as well and would ultimately involve "buying out such large chunks of Norco that you are really threatening the integrity of the town." In Shell's representation of the VPPP, then, offering to buy two streets promised the maximum benefit for the community: it created a buffer zone that would help insulate residential areas from the noise of plant operations and up-close views of operating units. At the same time, it left in place not only the bulk of Norco's population but also the town's commercial core (which lies within four streets of the east site fenceline), a central feature of the community's identity as a thriving small town.

Two years later, the Diamond Options Program reflected a similar preoccupation with community image in Norco, but underscored the centrality of private property. Where the VPPP had offered property owners an incentive to sell their homes to Shell, paying 30 percent over fair market value,[23] the Diamond Options Program offered homeowners on Diamond's two remaining streets an incentive to stay in Norco. If they chose not to accept the market value of their homes—in this case adjusted for the presence of the chemical plant—they could opt to take a home improvement loan of up to twenty-five thousand dollars. The "loans" would be interest-free and forgiven at a rate of 20 percent per year, meaning that they cost nothing for residents who remained in Diamond for an additional five years. The program's objectives, which state that Diamond Options is meant to "support and complement" the Good Neighbor Initiative, make sense of the home improvement aspect of the program in terms of community image in more generally. The program's third objective was to "maintain the desirability of certain neighborhoods to prospective buyers and tenants"—to, in other words, convey to outsiders that Norco was a nice place to live. To the extent that the home improvement loans enticed residents to remain in Diamond, they helped subvert the obvious message of a large buyout won through a protracted environmental justice campaign: that Diamond was too polluted a place to contemplate buying a home. At the same time, the loans helped ensure that Norco was seen as a nice place to live by ensuring that it consisted of well-kept properties. Brignac explained that the home improvement loans addressed "socioeconomic issues" in Diamond, namely, that "when you ride through Diamond it becomes immediately obvious that the homes are not as nice as in this [the white] part of Norco." Making money available for residents to fix up their homes thus helped protect Norco's status as a "nice little town" by making properties in Diamond more comparable to those in other neighborhoods.

Caring for Community in New Sarpy

While entrepreneurialism in Norco took the form of extensive community improvement efforts, New Sarpy saw few comparable coordinated efforts to build up the community—at least prior to the summer of 2002. With the exception of an after-school tutoring program run by the largest of the black churches in the community[24] and a brief mobilization among some black residents to build a playground, residents cared for their community by attending to their own properties and helping less well-off neighbors and relatives attend to theirs. Even amid the intense controversy that ignited with Orion's offer of the Community Improvement Program in July 2002, mowing lawns, washing siding, and keeping gardens remained central to the everyday life of homeowners, including CCNS leaders. When I visited Ida Mitchell to talk about the campaign, for example, she would often tell me how she had been up early to work in the yard before the heat of the day. Residents also periodically made small improvements to their properties: that summer Harlon Rushing added a canvas carport in his driveway, so he would have shelter for his truck; one of the Berteau households installed a basketball hoop for their children—and their nieces and nephews and other kids along the street.

The contribution of this work to the quality—and image—of the community, however, only became noticeable in the context of residents' complaints about their less responsible neighbors. When Irene Masters saw me to my car after our interview, she drew my attention to the property next door to hers. The house was boarded up, with construction waste piled in the yard. The family that owned the property had neglected it, Masters told me, and it was clear that she thought it was a disgrace. I had similar encounters with a number of other residents, who complained to me about the neighbor who did not take out his trash or the neighbor who would not mow her lawn or the ramshackle house on the corner whose porch seemed to be permanently occupied by a couple of drunken men.[25] Though they never said so, I believe that residents pointed out these eyesores to me in order to make sure I would not get the wrong impression of their community. By expressing their disapproval of the messy neighbor, they asserted that such neglect was outside the community's norm, that the boarded-up house was a blotch on an otherwise nice neighborhood.

These eyesore properties arguably had an effect on the community that went beyond aesthetics. Lamenting unkempt properties and widespread litter, Audrey Taylor explained it this way:

> I used to tell my kids when they was young and had company, I would say, look, if you don't straighten up your room and have it looking nice when your friends come in, they're going to throw things everywhere and it's going to be even worse than it looks now. I said, but if you go into a room where everything is in order, and you're playing with things, you're going to put it back, because you're going to want to leave it like it looks.

The same logic extended to the neighborhood, Taylor implied. Because some of her neighbors did not tend to their properties and failed to take care of their trash, passersby could think it was acceptable to throw trash in the drainage ditch in front of the Taylors' tidy home—adding an extra chore, raking out the ditch, to the work that the aging couple did to maintain their property. Keeping the place nice was easier when it was already seen as a nice place.

Complaints about messy neighbors in New Sarpy, then, underscore the importance of private property to the image of the community as a whole. In the absence of collective entrepreneurial work, community image was constituted by the aggregate condition of individual properties. Homeowners who worked to maintain their properties struggled to have the standard they set be seen as the defining norm of the community, and not have New Sarpy's image be colored by their less diligent neighbors. While this dynamic was especially visible in the absence of coordinated community improvement programs, it would probably not have been obviated by them. In Norco, where entrepreneurialism took more collective forms, Shell still saw value in giving loans to bring individual properties up to the level of the rest of the community, and NCA, for its part, focused a significant portion of its effort on cleaning up vacant and abandoned properties.

Insulting the Community

In New Sarpy, where maintaining the quality and image of the community depended on the upkeep of individual properties, each messy neighbor was an insult to the whole community. Their unkempt properties made the neighborhood look uncared for and undermined the work that other residents did to establish it as a nice community. At the same time, the presence of the refinery next door created other, larger insults. Orion was itself a messy neighbor on a grand scale, and even beyond its unsightly presence, its operations worked against the "nice" quality that community members were striving for. But residents' attempts to get the refinery to clean up also constituted a kind of insult. In campaigning for clean air and relocation, CCNS, like CCN in Norco, indirectly asserted that the refinery-adjacent town was not a nice place to live—an insult felt keenly by those less riled by Orion.

Insults of Industry

When Audrey Taylor showed me out of her house in March 2003, what she pointed out to me was not the mess next door. Instead, she drew my attention to her own house, showing me how the siding was coated unevenly in some sort of dark green substance. She said she thought that was new, that they did not get this stuff on their houses when she was growing up in New Sarpy. It was clear that she believed Orion to be somehow responsible for the green stuff. Gesturing to a dingy picket fence that ran along the side of her property, she told me that it used to be the same color as the snow-white azalea that bloomed in front of it. I looked closer to see that the fence was coated with the same substance that covered the house. Her husband really ought to get out and scrub it off, Taylor said, but with his heart condition, it was hard for him to do, and they had not yet found any neighborhood teenagers whom they felt they could trust with the job.

Down the street at the home of Harlon and Janelle Rushing, the story was similar. On my first visit to the Rushings in July of 2002, Harlon Rushing had been furious with Orion for sticky black dots that coated his house, and truck, and trees. He told me that he washed his truck every morning, only to have it quickly covered again in "soot," as he called the black stuff, and he pointed out the carport that he had put up to protect the truck from Orion's fallout. Looking at the dark green tarp stretched across an aluminum pipe frame, I was surprised to hear that the carport was only a month old: the soot deposits had already dirtied it and made it look weather-beaten.

These encounters pointed to a significant kind of impact that Orion had on quality of life in the neighboring community. Beyond any potential health problems it might cause, the refinery's pollution sullied property. It made white fences look gray, it made new structures look old, and it made the houses of lawn-mowing, ditch-raking, truck-washing homeowners look unkempt. The effect was distressing to residents like Taylor and Rushing in part because it undermined their efforts to keep their neighborhood nice by maintaining their properties. Caring for their aging homes was already a challenge for the two couples, both in their seventies, but deposits of soot and other substances from the refinery made it nearly impossible for them to keep their houses as clean as they would have liked.

But beyond making more work for careful homeowners, pollution from Orion affected the way the neighborhood might appear to outsiders. Outdoor structures that were beyond residents' ability to clean, or that would not stay clean because of Orion's continuing emissions, became visual clues to the town's character—hinting that perhaps the more overtly run-down properties scattered throughout the neighborhood were not so exceptional after all. Indeed, by pointing out her own dingy fence to me on my way out, Audrey Taylor was arguably making an effort to counteract that impression, in the same way others did by condemning their neighbors' messes. The less-than-white fence was certainly not up to *her* standards, she implied; it was simply out of her control.

The potential effect of refinery emissions on community image was underscored by New Sarpy resident Clarice Watson, an elderly black woman who lived within sight of Orion's huge storage tanks and constantly burning flare. Watson believed that toxic gases released by the refinery exacerbated her heart condition. In one of my visits to her double-wide trailer, she described the intensity of the chemical odors in her home when the refinery had a release: at times the air was so bad, she said, that she'd stick her head in the refrigerator just to get a breath of clean air. But then she went on to say how the smells would linger in the house for days after one of these incidents. As a result, she said, "People come to your house and think you don't clean." Here again, the refinery's pollution not only affected how comfortably, and safely, residents could live in their own homes. It also affected the perceptions of others, at least as imagined by residents: visitors who attributed lingering smells to Watson's poor housekeeping, like those who took Taylor's dingy fence as the norm for the neighborhood, were likely to leave with the impression that New Sarpy was not a clean or well-cared-for community.

Indeed, throughout most of CCNS's campaign, it seemed that Orion itself had such an impression: representatives of the refinery acknowledged neither the work that residents put into caring for their properties nor the importance they placed on being seen as a nice place to live. Their disregard for residents' concern with the appearance of the community was particularly obvious in an encounter Harlon Rushing had with an Orion employee in the summer of 2002. As Rushing told it, he had called the refinery to complain about the soot, and Orion agreed to send someone out. When their representative arrived, he made a cursory inspection and told Rushing that the soot was just dirt. Rushing recounted this pronouncement incredulously and went on to say how he had argued with the Orion representative. Showing him how the soot coated the top but not the bottom of the leaves on his fruit trees, and the hood but not the undercarriage of his truck, Rushing reminded him that "dirt don't just jump up on things." Rushing's reasoning did not persuade the man to take responsibility for the soot; he suggested that perhaps it was from the Shell facility in Norco and then drove away, leaving Rushing to continue to stew over his sooty property.

Rushing's frequent retellings of this incident emphasized the ridiculousness of the Orion representative's claim that the airborne substance coating the neighborhood was just dirt. But in the context of residents' concern with caring for the community, the man's statement was not just ridiculous; it was also an insult. Suggesting that the black deposits were something as innocuous and mundane as dirt, the refinery representative also suggested that it was something that Rushing could take care of himself—and, further, implied that he was trying to blame Orion for his own failure to keep the place clean. In this encounter, then, community image was centrally at stake. By advancing an image of New Sarpy as a dirty, uncared-for place, the Orion representative could deny that the refinery was affecting the quality of the community. Conversely, in pointing out the illogic of the man's statements, Rushing was not only asserting his understanding of basic physics but also challenging an outsider's allegation that he and his neighbors would not know how to deal with plain old dirt.

In their treatment of the section of their facility nearest the community, Orion showed a similar disregard for New Sarpy residents' understanding of their neighborhood as a nice place. The land between the homes on St. Charles Street and the row of large storage tanks, some two football-field lengths away, had long been owned by the refinery. Originally wooded, the property had been cleared of trees to make way for a coke conveyor unit, which was ultimately built elsewhere on the refinery's property after successful community opposition to the original proposal. For a time, the land was used as a parking lot for refinery employees, and traffic through the dirt lot kicked up dust and created problems for residents of St. Charles Street not unlike those created by the sooty deposits of 2002. When Orion took over the refinery in 1999, the company began to use that section of the property as a "drop yard"; that is, it became a dumping ground for pipes, fittings, and pieces of machinery that were old or worn out and being replaced.

The drop yard gave residents of St. Charles Street a new messy neighbor on a grand scale. On the west side of the street, residents' back yards were separated by a chain-link fence from a growing pile of rusty metal parts and broken-down machinery. Large new castoffs were often added, sometimes late at night, with an earth-shaking thud. As an unsightly installation at the neighborhood's borders, the drop yard represented an insult to the community's image—and residents' work to maintain it—comparable to Orion's attempt to pass off sooty deposits as "dirt." By disposing of its trash at the edge of the community, the company disregarded (or, perhaps more likely, failed to understand) the central role that tidy properties played in making New Sarpy a nice place to live, both in the minds of residents and in the eyes of outsiders. It seemed instead to suggest that a pile of industrial trash would not have a significant impact on the quality of the community—to imply, that is, that the community was trashy to begin with.

Unlike Shell and Motiva, which were explicitly and actively engaged in improving quality of life and supporting community entrepreneurialism in Norco, Orion was arguably undermining community quality and initiative in New Sarpy. But it was not Orion's mere presence, or even the views of the tanks and flare from the St. Charles Terrace neighborhood that made it seem a less nice place. Indeed, even having better vantage points on intricate operating units and looming stacks in Norco did not diminish Milton Cambre's, and others', view of it as a "nice town" or a "good community." Rather, Orion detracted from community quality and image with dirty, smelly emissions that overwhelmed even diligent homeowners and made the community appear less cared for than it really was. Even more importantly, the refinery itself seemed to cultivate a view of the community as dirty or unkempt by failing to acknowledge that its sooty emissions and industrial junkyard could bring down the community. As my conversations with Taylor, Rushing, and Watson suggest, these insults—while not overwhelming community concerns with health and safety issues—were nonetheless deeply felt by residents working to build a home and a neighborhood that others would recognize as a nice place.

Insults of Activism

New Sarpy residents talked about Orion's insults not only in everyday talk about their neighborhood but also in public statements made as part of the campaign for relocation. Indeed, CCNS members' frustration with Orion's foul smells and soot were central drivers of their campaign, both as insults in themselves and as perceptible markers of the damage residents imagined refinery emissions to be doing to their bodies. In the context of the campaign, however, residents' complaints about Orion's insults took a different valence than in everyday talk. Instead of casting the soot and drop yard as blotches on a nice community, campaign talk presented Orion's operations as fundamentally altering the character of the community. For example, *Land Sharks*, a report released by CCNS in conjunction with the Louisiana Bucket Brigade (LABB), quotes New Sarpy resident Shonda Lee—a middle-aged African American woman whose family is described in the report as having a "deep history with the land on which they live"[26]—testifying to the smells and other disruptions caused by the refinery:

> The smell. Yesterday was so disgusting, yesterday I was in the car . . . and the smell was so awful, we were sick to our stomachs. We left New Sarpy and felt much better. We got back here and we were sick again. This is no lie, sometimes the smell is so bad I hang out of my door and throw up.

While in this statement Lee describes the same odors that residents like Watson and Rushing complained of, she does not contrast them to the nice neighborhood the residents tried to maintain. Instead, she describes the odors as characteristic of her community: yesterday's events are offered as an example of how merely being in New Sarpy can make one ill.

Other parts of Lee's statement, as reported in *Land Sharks*, similarly paint a picture of New Sarpy as a rather grim place to live:

> It's at night, when we're sleeping. The flare burns. The rumbling, the noise. I hear it so clear at night . . . when we're really trying to sleep to get up for the next day. We lose a lot of sleep. . . .
>
> My daughter wakes up in the middle of the night because she's afraid. She's very fearful due to the big explosion [fourteen-hour tank fire] that just happened. She even had a nightmare the other night. She dreamed . . . that Orion just blew up.

In Lee's description, Orion is much more than a blotch on an otherwise lovely neighborhood. It is instead a looming danger whose effects pervade life in New Sarpy, intruding even on residents' sleep. The reader is left to wonder how residents can stand to live there at all.

Echoed by other residents quoted in the *Land Sharks* report, this characterization of New Sarpy as virtually unlivable was integral to the message of CCNS's campaign for relocation. Orion's operations—their flares, accidents, and emissions—had made New Sarpy a dangerous, illness-inducing place to live, CCNS members claimed in their public statements. As a result, they asserted, the company had an obligation to buy residents' homes at a price that would allow them to relocate to an area where they could feel safe and breathe easily. The call for relocation thus insisted that a "nice neighborhood" was only to be found somewhere else—in contrast to residents' (including CCNS members') representations of their community in other contexts. Though politically potent, this characterization of New Sarpy was an insult to the community in itself. In pressing their demand for relocation, CCNS members were saying that New Sarpy was not a nice place to live at all. While they had to make such a claim in order to justify their demand, in doing so they disregarded the everyday work residents put into keeping up their homes and the community in much the same way that Orion did when they put their drop yard right next to the neighborhood.

The use of technical information in the campaign extended the insult by making the case that New Sarpy was an inherently dangerous place to live. When bucket results were publicized, for example, they were presented as evidence of the poor air quality that plagued the community. In September 2000, CCNS presented Orion with results from a bucket air sample showing very high levels of benzene in the air. Speaking to reporters covering the event, resident Dorothy Jenkins said of the bucket results, "This is what's harming us all, these chemicals that we're living under." That is, the high chemical concentrations in the sample represented what it was like to live in New Sarpy—not what had come out of the refinery during an isolated incident.[27]

The *Land Sharks* report, which presented extensive data collected and compiled by CCNS members and LABB, similarly made the case that the refinery had made New Sarpy a polluted and dangerous place to live. Spread across several pages bearing the heading, "The Urgent Case for New Sarpy Relocation," Orion's accident reports, filed with the Louisiana Department of Environmental Quality (LDEQ), provided the basis for the report's claim that the refinery posed a consistent, unabating threat to residents through the flaring of sulfur dioxide (SO_2), constant accidents, and repeated equipment failures. A timeline of accidents from January until August 2001 spreads across two facing pages, showing leaks, spills, fires, and explosions in almost every month, with seven in June alone. The timeline excludes SO_2 flaring, which is summarized in a separate table, with columns showing how many pounds of SO_2 were released, in how many flaring incidents, for every month from May 2000 until July 2001. The table reports totals of nearly two million pounds released in 108 accidents, then breaks the releases down into monthly, weekly, and even daily averages (the last a whopping 4,356 pounds). Arrayed in this way, the data from accident reports paints a picture of New Sarpy as a town besieged by an accident-prone refinery, a neighborhood where people breathe SO_2 daily, a community where people imagine that an accident could cause "major disaster at any moment."

Similarly, the report presents results from bucket air samples taken by residents in a way that emphasizes the problem of pollution as endemic to the community. On the page following the accident timeline, a table summarizes chemical concentrations measured in six bucket samples taken between August 2000 and August 2001; however, the results are not organized by incident, as one might expect. Instead, they are organized by chemical: boxes entitled "Hydrogen Sulfide," "Carbon Disulfide," "Carbonyl Sulfide," and "Benzene" are arranged two by two. Each box includes a phrase describing the health effects of the chemical (e.g., "attacks the nervous system"), the Texas Screening Level for the chemical, the Louisiana state standard ("none" in all cases other than benzene, for which a Texas level is not included), and either three or four dates, each with the concentration measured on that date listed alongside. Organizing the bucket results in this manner downplays the individual releases that produced the high chemical concentrations; indeed, one must search out the unique dates in the table to see that it covers six releases, rather than four (the largest number of dates listed in one box) or fourteen (the total number of date-concentration pairs in all four boxes). Instead, the unusual presentation of data emphasizes the frequent presence of chemicals with known effects on human health—and, again, tells a story of New Sarpy as a hazardous place to live.

In their campaign for relocation, then, CCNS members not only declared their community unlivable but mustered quantitative data to prove it. That their rhetoric constituted an insult to the community that they worked so hard at keeping nice, and resented Orion so much for sullying, went unnoticed or at least unremarked for much of the campaign, even by residents who participated in both discourses. Yet the stakes were the same: to the extent that being seen by others as a nice place to live is central to the success of entrepreneurial communities, activists' insult had the same potential to damage New Sarpy as Orion's did. By publicizing at every opportunity their claims that New Sarpy was awash in pollution and endangered by industrial accidents, CCNS members risked tarring the community's image and driving away prospective investors—most immediately homebuyers seeking a good community for their families.

Building Up the Community

Activist insults to community image might seem not to matter, if residents were to be relocated anyway. But relocation was not CCNS's only objective. The campaign's full goal was "relocation for those who want it; clean air for those who don't," in the words of Ida Mitchell, who saw the Clean Air Act lawsuit against Orion as being at the heart of CCNS's strategy for achieving the latter. In its efforts to be inclusive, though, CCNS neglected the ways that its strategies worked against improving New Sarpy for those who wanted to stay. Advocating for reduced emissions, better safety, and more complete environmental reporting addressed Orion's insults, but calling the community unlivable and threatening to strip it of the substantial portion of inhabitants who did want to relocate created the possibility of real negative effects on both community quality and community image.

The tension between CCNS's campaign goals was glossed over—as the same tension had been during CCN's campaign in Norco—until July 2002, when Orion's offer of home improvement loans and cash payments for residents generated a competing discourse of community improvement in New Sarpy. For the next several months, CCNS members struggled with other New Sarpy residents over how best to "build up" the community, especially over whether Orion should be seen as an obstacle to or a partner in community improvement projects. During the same period, in Norco, white residents and remaining Diamond residents were working to reestablish Norco's identity and image as a good community in the aftermath of CCN's successful bid for relocation, a project in which Shell was a key collaborator. In both cases, positioning industrial facilities as partners in community improvement gave residents considerable incentive to accept the technical authority of industry experts—even on potentially contestable claims: in New Sarpy, taking for granted Orion engineers' ability to make environmental improvements allowed CCNS to take advantage of resources for improving the community that Orion offered to provide; in Norco, experts' air quality data helped recuperate a community image that had been tarnished by activist insults.

Community Improvement in New Sarpy

Even as its soot blanketed New Sarpy homes and its drop yard remained the messiest neighbor on St. Charles Street, Orion took up the mantle of community improvement. At a tense community meeting on July 17, 2002, Orion community relations manager Henry LeBoyd, a middle-aged black man who had grown up in New Sarpy and pastored one of the community's churches,[28] laid out the refinery's Community Improvement Program (CIP). Orion was offering each homeowner a loan of twenty-five thousand dollars—interest free, with the principal forgiven at a rate of 20 percent per year—to make improvements to his or her property. People who owned rental property were eligible for two loans: one for their own residence and a second for one of their rental units. If a homeowner chose not to take a home improvement loan, he or she could opt to receive a series of cash payments—fifteen hundred dollars every six months for five years, for a total of fifteen thousand dollars. These payments were also available to renters whose landlords chose not to take a loan on the properties they occupied. LeBoyd explained that the cash, which he dubbed the "Family Enhancement Plan," could allow residents to further the community in ways other than fixing up their houses: they could pay for continuing education, for example.

Modeled on Shell Chemical's Diamond Options Program, which offered the same home improvement loan option to residents who did not wish to relocate, Orion's CIP likewise built on the idea that improvements to individual properties constituted improvements to community quality; moreover, that logic was extended to whatever use residents might put cash payments to, as a way of justifying the Family Enhancement Plan as something more than paying people for their quiescence. But in his presentation, LeBoyd argued that his company's CIP did more to further the quality of the community than the Shell program on which it was modeled, because it did not include a relocation option. Orion, LeBoyd said, wanted to invest in the community, not see it gutted by people moving out.

While Orion's CIP in many ways resembled other industry-sponsored programs for community improvement, it departed from all of them in one important way. The CIP was conditional. It would only be implemented if and when CCNS dropped their Clean Air Act lawsuit against Orion.

CCNS members immediately jeered LeBoyd's proposal as yet another insult. At the meeting and in a press conference the next day, residents called the offer "chump change" and accused Orion of trying to dictate to the community.[29] Further, CCNS members rejected the idea that Orion's money would better the community, equating clean air, not home improvements, with community quality. During the press conference, speaker after speaker talked about the company's pollution: CCNS president Dorothy Jenkins began by describing how foul the air had been the two preceding nights; Shonda Lee asserted that the air in New Sarpy was worse than it had ever been in Diamond; Harlon Rushing complained of the soot; and Guy Landry talked about people who had died of cancer, especially brain tumors—implying that they were caused by pollution from the oil refinery. The message, variously articulated and implied, was that CCNS's campaign was fundamentally about clean air, and that Orion's CIP, in the words of Don Winston, "doesn't address the seminal problem: pollution." Worse, because the offer was contingent on CCNS dropping their lawsuit, Orion was asking residents to discontinue efforts to address the problem. As one elderly black woman put it, the company wanted them to get a little bit of money to live five more years in the same filth.

In addition to charging that Orion's CIP worked against community quality by failing to address pollution issues, CCNS members accused the company of trying to divide their community. At both the meeting with LeBoyd and the press conference the following day, residents took issue with the way that Orion had gone about proposing the CIP, going through the neighborhood with a petition that residents were asked to sign to indicate their support for the program, rather than taking the plan directly to community leaders. Jenkins, CCNS's president, testified at the press conference that she had first heard of the plan from a neighbor, because the Orion representatives going door to door with the petition had not even visited her. These tactics were evidence that, in Jenkins's words, Orion was giving them money to try to split them up. As further proof that Orion was attempting to divide the community, Jenkins revealed to the group that LeBoyd had called her and told her that, if CCNS dropped their lawsuit, she and several other CCNS leaders living right along the fenceline would be bought out. She made clear that she did not trust his offer; rather, she saw it as an attempt to get her and the others to betray the rest of the community rather than standing together with them.

In these early reactions to the CIP, then, community improvement was conceived as clean air, pursued by a united community. But a number of residents who had not been involved with CCNS's efforts soon mobilized around the CIP, forming a rival community group called the St. Charles Terrace Neighborhood Association (SCTNA). Borrowing heavily from the Norco Civic Association, SCTNA articulated a particular vision of community improvement that resonated with New Sarpy residents' everyday ways of caring for their community—but that figured Orion as a partner in, rather than an obstacle to, that project.

As its founders would later tell it,[30] SCTNA was first conceived at an event hosted by Orion on August 7, 2002, three weeks after LeBoyd's presentation to CCNS. Held at a local restaurant, the "seafood boil"[31] ostensibly aimed to inform residents about the CIP; however, to enter the dining room, attendees not only had to show proof of residency in the area covered by the program but also to sign a form indicating whether they would prefer to take a home improvement loan or cash payments. Residents active in CCNS did not attend; some said that members of the group were categorically barred, while others said that they stayed away because they were unwilling to sign the form. No one who did attend ever spoke to me about what happened during the event. It was right after the event, according to SCTNA president

Jeffrey Burnham, that several attendees congregated in the parking lot and came up with the idea for a neighborhood association.

The goals of SCTNA, as Burnham and other officers of the organization explained at the organization's first open membership meeting in November 2002,[32] was to improve the quality of life in the community for residents and businesses. Among the ideas for community improvement offered by SCTNA leaders at the meeting included starting a neighborhood watch program, installing speed bumps to address the problem of cars speeding through the neighborhood, picking up litter and beautifying the levee, and helping elderly residents with maintenance on their homes. SCTNA's vision of community improvement also addressed the problem of messy neighbors: in a subsequent interview, Burnham explained,

> There's a lot of room for this area to grow as far as, you know, there are some sections where they got people living in trailers without power and trash-filled yards and stuff like that, and basically we just want to improve it, clean it up, you know, just improve the neighborhood.

Agnes Boudreau,* an older white woman who served on the board of SCTNA, summed up the purpose by saying that the new organization represented the fact that they wanted to stay in New Sarpy, beautify the community, and build it up for everyone.

Like CCNS members, who criticized Orion for trying to divide the community, SCTNA leaders invoked an ideal of community cohesiveness in their vision for the organization. Part of what the organization wanted to promote, Burnham explained both in the meeting and in my interview with him, was sociability among residents: he wanted to be able to walk down his street and call his neighbors by name. At the meeting, another board member suggested that SCTNA might operate like the neighborhood association in her parents' community, where decorating for holidays "really brought the community together, more so than they had been before." But unlike CCNS, which did not claim that everyone wanted relocation, merely that everyone should have the option, SCTNA leaders envisioned the group as a representative body that would speak with one voice for the community as a whole. In the meeting, Burnham explained the organization's representative scheme: once the group got off the ground, a general election would be held to select officers; in addition, each street would choose its own delegate to the board.[33] The group aimed to be representative and accountable, Boudreau said, and later in the meeting, she presented SCTNA as an advancement over CCNS: while she applauded what CCNS had done, she said, she felt that the community now needed to find its voice, to decide what it really wanted.

In their insistence on speaking with one voice and building up the community for everyone, SCTNA leaders implicitly challenged the logic of CCNS's campaign. CCNS's two-pronged goal of "relocation for those who want it; clean air for those who don't" implied that residents could simultaneously improve the quality of the community by addressing Orion's pollution *and* create the opportunity for residents to leave by persuading Orion to create a buyout program. For SCTNA leaders, in contrast, community quality hinged on residents' staying and working together to "build up" the neighborhood. The specter that loomed for them—and for anyone who was not eager to leave New Sarpy—was that if relocation were offered as an option, the community would disband. Those who stayed might be among only two or three families living on a given block; there would be no neighbors to nod to, to share home maintenance with, to make common cause with. Relocation could not be compatible with community improvement for residents who stayed, because it would effectively ensure the end of the community—a fact that CCNS's win-win campaign goal failed to acknowledge.[34]

Further, in SCTNA's vision of community improvement, Orion was conceived not as an obstacle, as CCNS saw it, but as a resource. The organization's founders maintained that, although they first decided to form a neighborhood association after Orion's seafood boil, the idea had been solely their own and not (as CCNS members alleged) suggested to them by company representatives.[35] But as they tried to organize themselves, they found that they had no idea how to go about it. According to Burnham, the group spent their first few meetings "stumbling around, staring at each other." It was only at that point, SCTNA leaders said, that they approached Orion for help. Orion responded by hiring a consultant to work with residents. The consultant, a local realtor with experience setting up neighborhood associations, obtained bylaws from the Norco Civic Association for residents to use as a model and prepared the paperwork necessary for SCTNA to incorporate as a tax-exempt nonprofit organization. Invitations to the group's first general membership meeting were also apparently sent out with Orion's help: the mailing list included, for a few residents, addresses that only the company would have had on file, and SCTNA leaders evaded questions about who paid for postage, saying only that the Orion-sponsored consultant had handled it.

Even assuming that the new neighborhood association was entirely the idea of community members, the fact that they so quickly approached the neighboring refinery for support is telling. Both Norco and the Ormond subdivision in nearby Destrehan had comparable organizations to whom SCTNA's founders might have looked for guidance; had none of them been acquainted with members of those organizations, which would have been highly unlikely, they might have enlisted the parents whose neighborhood association had helped inspire their own efforts. That they instead turned to Orion as their primary source of support in founding the organization underscores the central role afforded to industry in community improvement. Orion had, at least, expressed their desire to help build up the community through their CIP proposal; LeBoyd's presentation of the proposal touted other community outreach, as well, including an after-school tutoring program and contributions to the local United Way. Moreover, in approaching Orion for help, New Sarpy residents followed the lead of myriad St. Charles Parish community organizations, ranging from historical societies to Little League teams, who routinely appealed to local industrial facilities for support.

Similarly, although SCTNA leaders at the November membership meeting stressed that Orion would not run the organization, they fell back on the idea of industrial patronage when asked where they would find funding for the many projects they envisioned. There were state and federal grants that they could apply for, and they would also be accepting donations, Burnham said, without specifying who might be in a position to make charitable contributions to the organization. Money expected to come to the community as a result of enforcement actions against Orion was also cited as a potential source of funds for SCTNA activities.

CCNS's and SCTNA's divergent versions of community improvement—clean air and relocation versus unified, industry-supported beautification, respectively—came into conflict most directly during an "informal meeting" with the LDEQ about the agency's enforcement action against, and settlement with, Orion. Announced in early September, the LDEQ's settlement with Orion had been a major blow to CCNS. Their lawsuit against Orion, which alleged that the refinery had violated the federal Clean Air Act (CAA) on a number of counts,[36] was predicated on the assertion that the government agencies responsible for enforcing the law had not prosecuted the company for its violations. In such a situation, the Clean Air Act explicitly grants citizens the right to act as enforcers; CCNS thus had standing to bring its lawsuit, filed in November 2001, because neither the LDEQ nor the U.S. Environmental Protection Agency (EPA) had cited Orion for the violations alleged in the suit. When in August 2001, the EPA served Orion with a Notice of Violation—telling the refinery, in effect, that the agency found them to be out of compliance with the CAA on the points that CCNS's lawsuit covered—CCNS and their counsel at the Tulane Environmental Law Clinic cheered: the EPA had just made it much easier for a judge to agree that Orion had violated the CAA. A few weeks later, though, the LDEQ announced that Orion agreed to pay one million dollars in fines and spend two million dollars on "beneficial environmental projects" (BEPs), in addition to taking action to address excessive emissions, poor reporting, and failure to monitor. The LDEQ-Orion settlement made it far, far harder for CCNS's lawyers to continue to argue that government agencies were not enforcing the law, and created the distinct possibility that a judge would rule that CCNS did not have standing to sue Orion, ending their lawsuit.

The settlement thus jeopardized the biggest weapon in CCNS's campaign: not only did the group hope it would win them clean air, but they also counted on the publicity and pressure they were able to generate through the ongoing suit to advance their demands for relocation. Hoping to challenge the settlement's adequacy, CCNS leaders requested a public hearing. The LDEQ refused, granting them instead an informal meeting. The difference, LDEQ representative Harvey Behler* explained at the meeting, was that there was no stenographer to record a transcript of residents' comments, and regulators would have a chance to not only hear but also respond to residents' concerns, which they did not really have an opportunity to do in public hearings with a formal process. CCNS was also asked, informally, prior to the meeting, to limit the number of attendees from the group to seven, a request that could not have been made for a public hearing.

At the meeting, most CCNS members commented on Orion's insults, positioning the refinery and its pollution as the primary obstacle to community quality. The soot and oil droplets that rained down on the neighborhood were a particular focus: Clarice Watson told regulators that she would no longer let her children eat the fruit from her fruit trees because of all the black stuff that coated it; Betty Morales, a white woman in her thirties, complained that an above-ground pool she set up had been similarly coated, making her loath to let her three small children swim in it because, as she said, "chlorine will only clean so much." She went on to express resentment that so much of her time off was spent cleaning stuff, yet her house still looked a disgrace. Ida Mitchell responded that it was a waste of time to clean the sulfur off her car; it just got right back on again. "I shouldn't have to want a buyout," said Morales, who grew up in New Sarpy—implying, again, that Orion had made conditions unlivable even for those who would prefer to stay.

In principle, the pollution issues that CCNS members complained about would be taken care of under the LDEQ-Orion settlement. Indeed, on hearing tales of the soot, Behler told the residents that his inspectors did not agree with the way Orion had been operating; it was their disapproval, one would infer, that led to the enforcement action, which aimed to bring the refinery's emissions down to acceptable levels. But in CCNS members' responses to the settlement, they expressed doubt that Orion could be trusted to make the necessary corrections. Orion's compliance would be verified by the LDEQ, Behler said in reply to a question from Don Winston; there were also penalties written into the settlement in case Orion did not comply. But his reassurances only led to further questions from CCNS members, who alleged that the LDEQ had known that there were problems even when they issued Orion's permit and had failed to crack down on the refinery either then or at any point in the intervening years. Ida Mitchell told the regulator flatly that she did not believe him or the Orion representatives lining the room: "We've been lied to so many times."

CCNS members' statements at the meeting on the settlement thus cast Orion as incompetent, duplicitous, or both—in any case an obstacle to community quality that could not be counted on to change its ways. In contrast, the settlement itself afforded Orion a central role in community improvement. Besides the measures the refinery would take to mitigate its air pollution, Orion agreed to spend one million dollars of the money set aside for BEPs on what Behler called "community-wide projects." While those projects were "loosely defined" in the settlement—in order, he said, to give Orion flexibility to respond to potential changes in what the community needed—he suggested that they would include an ambient air monitoring program similar to Shell's "Air Monitoring...Norco" program, a new early alert system, an enlargement of the "buffer zone" between Orion and the St. Charles Terrace neighborhood, and other, as-yet-unspecified projects to "enhance the overall quality of life."

The settlement agreement that Orion reached with regulators thus codified and sanctioned a vision of community improvement in which the refinery was not an obstacle to but a provider of "enhancements" to quality of life, including green space and trees in the space between Orion's storage tanks and the fence that separated the refinery from the neighborhood. By building flexibility into the BEPs, the settlement also figured Orion as a potential partner in defining and responding to community needs as they arose. This vision of Orion as partner and provider reinforced SCTNA's model of community improvement, and board members Burnham and Bourdreau, also present at the meeting, were quick to commend the settlement. Presenting a letter from SCTNA—which an attorney from LDEQ promptly offered to place in the settlement record—Burnham told regulators that he was satisfied with the changes that Orion wanted to make and happy with what they wanted to do for the community; he was, as he repeated numerous times, "in total agreement with the settlement." Boudreau suggested that the settlement would both keep Orion accountable and help the neighborhood improve; in fact, she said that she hoped that New Sarpy would set an example for other communities, giving them a vision of how they could work together with an industrial neighbor, rather than seeing their community disband as Diamond had.

In the positioning of Orion as a contributor to community improvement, it is important to note, the technical competence of refinery personnel was taken for granted. In the informal meeting, CCNS members' attempts to challenge the settlement on technical grounds were summarily dismissed. When Don Winston asked whether the settlement would really improve air quality, for example, Behler did not say, "We think it will, but we will look to the ambient air monitoring program to verify that it has." Instead, he declared baldly that it would: Orion would be upgrading to low-emissions technology. When Winston questioned the value of the ambient-air-monitoring program, suggesting that the data would not be useful without health monitoring, the regulator defended the approach of refinery experts. Telling Winston curtly that he was missing the distinction between emissions monitoring and ambient monitoring, Behler insisted that the ambient air monitoring included in the settlement was precisely the kind of data that health professionals could use, and that it would be necessary as a first step to any kind of health research that Winston might be envisioning. SCTNA members' support, too, reinforced the technical authority and competence of refinery engineers, with Bourdreau claiming that air quality had already improved since Orion took over the facility.

With the assumption that Orion engineers could reform operations at the refinery and produce clean air in New Sarpy written into the settlement agreement, it became very difficult for CCNS members to continue to question Orion's technical competence. Their contention that the company was unable or unwilling to run the refinery in a way that allowed them to breathe easily in their own homes was met by additional cadres of experts—regulators at the LDEQ *and* regulators at the EPA, who made clear that they also supported the agreement—who both took for granted and, in effect, guaranteed Orion's ability to improve its environmental performance.[37]

As experts closed ranks around the settlement, CCNS members came under increasing pressure to drop their CAA lawsuit and accept Orion as a partner in community improvement. The informal meeting made it clear that CCNS would be unable to stop or alter the LDEQ-Orion settlement, and CCNS leaders became pessimistic about the likelihood that their lawsuit would be allowed to proceed. At the same time, Orion began threatening to withdraw their offer of the CIP if CCNS did not agree to drop or settle their lawsuit before the settlement with the LDEQ was finalized. CCNS faced the possibility that both the home improvement loans and the lawsuit, which they saw as the lynchpin of their campaign, would evaporate. Moreover, even if they somehow managed to win their suit, the BEP money would still be the only direct benefit the community would see: any additional penalties ordered by the courts would be paid to the state treasury.

With their hard work threatening to come to nothing and neighbors beating on their doors at all hours to demand they drop the lawsuit, CCNS leaders began to pursue the model of community improvement espoused by SCTNA, codified in the Orion-LDEQ settlement agreement, and, indeed, integral to residents' everyday practices of maintaining community quality. At the beginning of November, CCNS leaders arranged a meeting with Orion's plant manager, deliberately excluding their Tulane Environmental Law Clinic (TELC) lawyers and their LABB supporters. They went with, in Don Winston's words, "a wish list" of things that Orion should do that would really improve the community. Topping the list was a major modification to Orion's proposed CIP: CCNS leaders thought that the maximum amount for home improvement loans should be doubled (to fifty thousand dollars) and the cash payment option should be eliminated in order to better promote real improvement in the neighborhood. The wish list also included measures that would make Orion into less of a messy neighbor: residents wanted them to pick up the drop yard, repaint the storage tanks to make them blend into the landscape, plant greenery to obstruct residents' view of the tanks, and landscape the area just inside the refinery fenceline along River Road. Finally, CCNS leaders asked for things that would improve the community as a whole: they wanted Orion to provide new street signs, to mark the community with a flowerbed, flagpole, and commemorative sign on the levee, and to provide assistance with home maintenance to senior citizens in the neighborhood. While Orion refused to eliminate the cash payments or increase the amount of home improvement loans in the CIP, the company agreed to many of the community-wide improvement projects and improvements to its own property—helping to satisfy CCNS leaders that company officials were sincere about wanting to build up the community.

The presumption of Orion's expertise in the LDEQ settlement thus pushed CCNS leaders to approach the company as a partner in community improvement, just as SCTNA had: with regulators lined up behind Orion's engineers and scientists, continuing to challenge the refinery's technical authority became quixotic, if not impossible. But at the same time, CCNS members' decision to fashion a partnership with their erstwhile foe provided significant incentive for them to accept Orion's technical authority. On December 18, community members met to vote on settling the lawsuit and accepting a slightly modified, formalized version of the CIP. CCNS's TELC lawyer explained that the LDEQ settlement covered much of the same ground as the CAA lawsuit, and that regulators whom they trusted also believed it to be a good settlement. CCNS leaders subsequently justified their desire to settle the lawsuit in terms that reinforced refinery and regulators' expertise: they had won clean air; the LDEQ settlement guaranteed it. The CIP, which they had in writing for the first time, simply provided additional benefit to the community. No mention was made of Orion's pollution, of the refinery's long-standing record of excessive flaring and accidents, of residents' lack of trust in refinery managers. To accept Orion as a partner in community improvement, CCNS leaders had to take for granted that the refinery could and would minimize its emissions and provide clean air.

Restoring Norco's Image

Accepting refinery expertise, which would have been considered questionable under other circumstances, thus became integral to the project of community improvement in New Sarpy. Only by conceding that Orion knew how to operate a refinery cleanly could CCNS take advantage of the company's "investment" in the community and partner with them to improve community quality. In Norco, industry expertise served a second, equally important role in building up the community: the town's image as a good place to live was restored, in part, by industry claims about air quality that might otherwise have been contested.

Both during and after CCN's campaign against Shell, whites in Norco expressed deep resentment of Diamond residents' activism. They accused CCN of lying about Shell's effects on their health, of campaigning for relocation just to get a little extra money for their houses, and of asking for special treatment that they did not deserve.[38] But one of the things that appeared to rankle white Norco residents most about CCN's campaign was the negative publicity that it brought their town. Like CCNS members, Diamond residents sought to show that their town was unlivable—that their air was fouled by Shell's emissions, that they lived in constant fear of the next accident or explosion at the facility, that the company had a long history of discrimination against and disrespect for its African American neighbors. But unlike CCNS, CCN succeeded in attracting a relatively large amount of attention to their claims. In addition to frequent news coverage in local and regional media outlets, including the New Orleans daily *Times-Picayune*, filmmaker Slawomir Grunberg made a documentary featuring CCN's campaign, *Fenceline: A Company Town Divided*, that debuted on PBS in 2002. Also in 2002, Steve Lerner, research director at the environmental nonprofit Commonweal, did a series of oral history interviews with Diamond residents and others involved in the controversy over their relocation. The oral histories were made available online soon after they were conducted to bring attention to CCN's ongoing campaign, and they were later synthesized into a book published by MIT Press.[39]

While the publicity was arguably crucial to CCN's success, media coverage representing Diamond residents' view of their town was, in the minds of white Norco residents, inaccurate and harmful. Talking to me after a Norco Civic Association meeting, one board member offered *Fenceline* as an example of the bad publicity that CCN had brought on. The documentary's representation of the town was skewed, the thirty-something white woman complained: it did not show any of the good things about Norco, including that it was actually an affluent place. Her concerns extended to the way environmental quality and health in Norco were represented in the press: she mentioned the moniker "Cancer Alley" as part of the bad publicity suffered by Norco and the rest of the heavily industrialized region between New Orleans and Baton Rouge, and she expressed frustration with CCN and LABB for saying publicly that the air was bad without being willing to show her the proof.

Residents who felt that Norco was being portrayed unfairly took every opportunity to defend their town. *Fenceline* in fact includes portions of interviews with a number of white residents who are proud of their town. In one, a white man seemingly in his thirties stands in front of a large, two-story home of the sort one finds in affluent suburban subdivisions. He tells the filmmaker that the nearby plant does not bother him: "I can think of a lot worse neighbors than an oil refinery." NCA's president, Sal DiGirolamo, is similarly featured, challenging activists' assertions that Shell's emissions are harmful by listing his octogenarian neighbors and contending that he could not think of any cases of asthma or cancer among all the Norco residents he knew. DiGirolamo defended Norco in his interview with Steve Lerner, as well, describing a set of interconnected factors that contributed to quality of life in Norco:

I think we have a relatively low crime rate. I think the happiness we enjoy adds to that. I know my neighbor, he knows his . . . the atmosphere adds to those things. I think it helps you to [live] a little longer life. All those things together makes NORCO a good town.

In conjunction with declarations about how he scarcely noticed Shell's noise and flaring, statements like this one by DiGirolamo and others aimed to paint Norco as a peaceful, pleasant place to live—and not the toxic tinderbox that CCN alleged.

The same Norco residents who defended their town during CCN's campaign were still occupied with Norco's image some seven months after CCN and Shell reached agreement on a relocation plan. The January 2003 agenda of NCA includes "publicity," for which a lead person is to be selected. The publicity person, the agenda states, will

> every month write an article on one or two items that we have accomplished or are pursuing and send to local newspapers. e.g. crape myrtle planting, banners, restrooms, Airline lights, Christmas lights, cleanup days, new home, etc. This type of publicity will counteract the negative press we get and show that our Norco is a good place to live and work.

The final line of the agenda item, which appears amid a host of items referring to activities that range from setting up charitable benefit events to getting new local ordinances passed, suggests both how important community image was to NCA's understanding of how to improve the community, and how threatened organization leaders believed it to be. An editorial in the local newspaper a month later lauded NCA's efforts to improve the community—and indirectly echoed their concern about bad publicity. Members of NCA, it said, "have been working for some years making improvements to an area which sometimes doesn't get noticed until a mishap at one of the refineries."[40] Commending the people of Norco as the "community's greatest asset," the editorial joined NCA's leaders in bemoaning the fact that the negative aspects of the community get all the attention—and contributed, whether wittingly or not, to the organization's attempts to counteract the trend.

Just as bucket data and other technical information had been important to CCN's and CCNS's efforts to show that industrial pollution had made their communities unlivable, technical claims were integral to white Norco residents' representations of their community as a good town. In the latter, however, residents relied on—and reinforced—the technical authority of industry experts. For example, where CCNS had condemned and even sued Orion for their excessive flaring, many Norco residents defended the practice in terms that echoed industry's justifications. Flaring, industry engineers frequently explained, was a safety measure. When there was some problem in their process and a unit went offline, the chemicals feeding that unit had nowhere to go. Letting them back up created a risk of a leak or explosion, so instead they would be redirected to a smokestack and burnt, obviating the risk. According to industry engineers, this practice of flaring was undesirable but innocuous: it wasted product, but the flares destroyed potentially hazardous chemicals so that they did not pose a threat to the surrounding area. Community groups like CCNS and their environmental allies charged that flares were so frequent that they looked more like standard operating procedure than a last resort; they also disputed industry experts' assertions that flares burned cleanly, without creating hazardous emissions.[41] But residents in Norco defensive of their town's image drew on industry experts' claims to declare flaring harmless. Speaking to Steve Lerner in 2002, for example, Sal DiGirolamo said,

> This is just an occasional deal when they shut down a unit. The safest thing you are going to do is to let that out. The safest thing to do is flare it. If you burn it, it goes away . . . not perfect but it is one of the best ways to get rid of that.

For DiGirolamo, flaring was not only an important safety measure; it also posed no threat to community quality: "We have this flare . . . one of these big units that shut down. . . . I took my granddaughter out there and said: 'Look at it.' She was not even interested [because she is used to that]. It doesn't affect my quality of life." Even during CCN's campaign, then, DiGirolamo and other white residents were using the technical claims of industry experts in their defense of their community. After the buyout was underway, they looked to Shell engineers to help undo the damage done to community image by CCN's campaign, especially their air monitoring with buckets, by representing environmental quality in a more positive light. In January 2003, Shell hosted a community meeting to update Norco residents on projects going on under their Good Neighbor Initiative (GNI), including "Air Monitoring...Norco" (AMN). The meeting was held at Norco Elementary School at the suggestion of a black resident, who reasoned that all segments of the community would feel comfortable there because everyone's kids went there, yet only white residents attended (Shell's contingent included one black community relations staff member). Although normally supportive of Shell, the crowd was surly: in discussions of the company's plans to turn land acquired in Diamond into a "greenbelt" and of a feature in the GNI newsletter on a relocated Diamond family, attendees expressed a deep resentment of CCN's campaign and anger at how Shell had given in to their demands.[42]

In the segment of the meeting devoted to AMN, residents pushed Shell representatives, at first subtly and then more overtly, to discredit data collected by CCN members. The question-and-answer period opened with a resident asking what the difference was between what Shell was doing in AMN and what CCN's bucket brigade had done. Taking over from the junior staff member who had presented the program, David Brignac, GNI's chemical engineer manager, answered that the difference was primarily one of equipment: AMN collected air samples in Summa canisters, while buckets collected samples in a plastic bag; the lab analysis was the same. He was pressed to put a finer point on the distinction later, when a woman asked, "The—what did you call it—canister, I'm sure must be far more superior to the ones used in the past, the ones used by Margie Richard and her group [CCN]." Shell representatives in fact refused to declare their technology superior—one of them responded by repeating bucket activists' claims that theirs is an "EPA-approved" method—instead offering that they would be getting much more data with the canisters than the buckets had collected.

Although Shell representatives were loath to discredit the buckets overtly, white residents' desire for them to do so pointed to their eagerness to distance themselves from unfavorable representations of the community—namely, bucket samples that apparently showed poor air quality. One resident even asked whether Shell would be doing a direct comparison between its data and results of the bucket samples that had been taken in the years prior, clearly hoping that the company could produce a picture of air quality that would supersede the one painted by CCN with their bucket results.

Norco residents' desire for a positive representation of Norco's air quality was eventually satisfied by results from Phase 1 of AMN, whose key findings were that Norco's air quality met state standards and that Norco's air quality was similar to that of other "cities."[43] The presentation of the results to the Norco-New Sarpy Community-Industry Panel in April 2003 made it clear that Shell and Motiva engineers had not fudged the data and were not simply pandering to community members. Not only had they worked with both the LDEQ and a small group of residents to develop a technically sound, credible monitoring protocol; they also resisted attempts in the question-and-answer period to absolve the plants of responsibility for the pollution that the monitoring program had found. When a parish councilman asked whether high levels of ethylene—a chemical that is produced, for example, by ripening fruit—might not be due to natural causes, Randy Armstrong chuckled and answered that, since Norco did not have any orchid farms but did have an ethylene plant, they could be pretty sure that Shell was the source. At the same time, industry experts' approach was not beyond critique. AMN adopted a sampling protocol standard in regulatory agencies but criticized by bucket activists for averaging out spikes in pollution; had an oppositional community group still been active in Norco, they would undoubtedly have been calling the program technologically backward and urging continuous, real-time monitoring at Shell's fenceline instead.

These criticisms did not emerge in part because the study's conclusions helped residents who remained in Norco reclaim an image of their town as a good place to live.[44] Expert representations of Norco's air quality put data behind their assertions that there was nothing wrong with the air after all; that flaring did not affect quality of life; that no one could reasonably be said to be dying young as a result of pollution. In fact, even the comparison of Norco's air quality to that in cities like Minneapolis and Houston makes more sense in this context: it flattered the small town by putting Norco in a league with much bigger, richer places. Community members' interest in creating a favorable image for Norco, in turn, helped restore the challenged authority of Shell scientists and engineers: their ready acceptance and uptake of AMN data helped establish the study's findings as facts,[45] and positioned industry experts as fact producers—a role that CCN had been unwilling to grant them.

Conclusion

On a gray January day in 2003, I went jogging on the bike path atop the levee near my apartment in Destrehan, hoping to run off my irritation with my boyfriend, visiting from Berkeley. I was soothed by the green slopes, by the view of the river, and by the steam billowing up from behind Orion's coker unit. Contemplating the unexpected beauty of the industrial landscape, I finally put my finger on what had me so annoyed. Every time we returned to my apartment from an excursion to New Orleans, my boyfriend would open the car door and pronounce, "They're not kidding it smells here" or "There's that smell again." He meant to be sympathetic—as did the environmental justice activists who told me I was brave for living "in the belly of the beast," and my advisor who wrinkled her nose and worried for my health when I told her where I would be living. But his suggestions that where I was living was smelly and toxic were also an affront: that was my *home* that he was talking about.

In that moment, I had my first glimpse of what it must have cost residents of New Sarpy to declare repeatedly and publicly that their neighborhood was not fit to live in—and, worse, to endure the sympathetic outrage of their activist allies, of visiting grant makers, of people like me, whose focus on how the community was being brought down by neighboring industry overshadowed their ability to understand it simultaneously as a home that people cared about and invested energy into making livable.

My boyfriend's expressions of sympathy injured my pride, a little. But for residents of New Sarpy, much more than pride was at stake. The dwindling of central government support for local programs and social services put the burden of entrepreneurialism on the community. That is, in a neoliberal age, ensuring and improving quality of life in New Sarpy and other small towns meant seeking out investors—including private-sector partners and responsible, upwardly mobile people willing to buy homes in the community—and competing with other small towns for their largesse, as well as for grant monies from the public and nonprofit sectors. Painting the problems in the community too starkly, or too publicly, created the risk of driving away potential home buyers, of making companies think twice about locating new facilities there lest they repel high-quality workers, of convincing established businesses and government agencies that their money was better spent in places where the problems were not already irremediable.

The need to be entrepreneurial led New Sarpy and Norco residents to imagine their industrial neighbors as natural partners in community improvement. Indeed, white New Sarpy residents occasionally expressed envy at how much Shell did for Norco, in contrast to Orion and its predecessors' neglect of their community. But the pressure felt by communities under neoliberalism to secure investment and put forth images as nice places to live also pushed residents to sideline their critiques of petrochemical facility experts: building up New Sarpy required accepting that Orion engineers could maintain clean air; defending Norco's "good town" image was aided by references to representations of Norco's air quality produced by industry scientists and engineers. Whether petrochemical industry experts were deserving of residents' restored respect is an open question: since CCNS's campaign ended, environmental conditions in New Sarpy seem to have improved a great deal; on the other hand, environmentalists could surely point to any number of ways they could be doing better still.

Regardless of the specifics of industry's environmental performance in any particular community, neoliberal demands on communities to be entrepreneurial push back against environmental justice critiques of expertise. They discourage the production of damning evidence about environmental quality or community health in fenceline neighborhoods; they afford industry scientists' data-backed denials of adverse impacts a role in building up the image of fenceline communities as "nice towns"; and they tie petrochemical plants' investment in communities to the assumption that plant engineers are capable of, and committed to, operating their facilities in a way that has minimal impact on residential neighbors. Operating on neoliberal terrain where community entrepreneurialism is both a virtue and a necessity, petrochemical industry experts rest their authority not only on the strength of their scientific knowledge and technical competence but on the ways in which their technical practices help "build up" fenceline communities and their image in the eyes of others—and thus overcome challenges from citizen science that, by raising questions about the livability of the neighborhoods nearest industrial facilities, threaten the image of would-be entrepreneurial communities.

4

From Deliberation to Dialogue

We are human beings and will be treated as equals.
—CCNS in *St. Charles Herald-Guide*, July 24, 2002

Nothing so destroys the sense of equality on which all pleasant social life depends as perpetual reminders that one member of the company out-ranks all the rest. When it is so, it is considered good manners for the pre-eminent one to keep quiet about it.
—Robertson Davies, *World of Wonders*, 1976

When Orion first announced its Community Improvement Program, members of Concerned Citizens of New Sarpy (CCNS) saw the company as trying not only to divide the community but to "dictate to" it as well, by setting out terms without first consulting residents. The community group tried to change the nature of the interaction by calling for talks with refinery officials: on July 24, 2002, the week after Orion's initial presentation of the program, CCNS ran a full-page ad in the local newspaper, inviting Orion CEO Clark Johnson to

discussions about a REAL Community Improvement Plan including
- Relocation
- Pollution reduction including sulfur dioxide and hydrogen sulfide
- Installation of enclosed ground flare system
- Reduction of coke dust in the neighborhood
- Clean up for the drop yard (promised by Mr. Johnson one year ago)
- Installation of perimeter air monitoring system

- Health clinic to deal with pollution effects
  . . . and other topics

The ad presented CCNS's campaign goals as representing true community improvement, in contrast to Orion's proposals, reflecting CCNS members' particular idea of what it meant to build up the community. Yet it also framed them as subjects for discussion and, further, suggested that it would be through respectful conversation that residents and the refinery would resolve their differences: "Talking face to face brings understanding," the ad read. "We are human beings and will be treated as equals."

CCNS members were not the only ones calling for talks—and in the process expressing the belief that discussions between CCNS and Orion were necessary to ending the conflict. Just a few days before their ad ran, an editorial in the same newspaper lamented that "the two parties are not at a point where they can speak to one another in a civil manner,"[1] suggesting that civility was the first step toward resolving hostility. Orion officials would later tell me that during that period all they wanted to do was talk to the community and that they believed that much of the trouble could have been avoided if they had been able to do so earlier. And the Louisiana Department of Environmental Quality (LDEQ) had already tried to facilitate conversation between the feuding parties by establishing in Norco and New Sarpy its first Community-Industry Panel—"a mechanism whereby community residents and industry officials can meet voluntarily as equals to discuss issues of concern" that would become a standard part of the agency's approach to addressing conflict between industrial facilities and residents of communities all over the state.[2]

With regulators, industry officials, and even local newspaper editors pushing the community toward talks, CCNS's invitation to Clark Johnson might have been little more than a strategic gesture—a superficial repackaging of campaign demands in the locally popular language of discussion. Yet CCNS members were actually quite sincere: they longed to be able to settle their differences with Orion through less conflictual means than marches and lawsuits. Their ideal, as expressed by CCNS leader Don Winston, was to be able to sit down with Orion's top decision makers and talk "like reasonable businesspeople."

The various calls for talks in New Sarpy evoked an idea powerful in American democracy—that problems can be solved, just laws made, and effective policies developed through deliberation. In the deliberative ideal, people come to discussion as equals; they set aside narrow personal interests to try to come to a mutual understanding of the common good; and they make use of reason and reasoned arguments in their exploration of public issues.[3] Elements of the ideal clearly informed residents' and regulators' visions of community-industry discussion. Statements by both groups, notably, emphasized that residents and industry must meet "as equals," underscoring the egalitarianism demanded by deliberation. They also used the language of common good and common concern: CCNS's ad frames the group's campaign demands as essential parts of any discussion about how to make the community better, while the LDEQ's description of its Community-Industry Panels refers to "issues of concern" in which all parties presumably share an interest. Finally, calls for talks imagine participants adopting a "reasonable" and "civil" tone—both preconditions of deliberation.

Yet beyond invoking the ideal of reasoned, egalitarian discussion of the common good, the visions of deliberation held by CCNS and their supporters, on the one hand, and the LDEQ and industry officials, on the other, were actually quite different. Through its ad, CCNS called for talks in which their campaign goals would be explicitly acknowledged as subjects of discussion; talks that would be facilitated by a third party committed to making sure community members were heard; and talks in which LABB's Anne Rolfes, lawyers from the Tulane Environmental Law Clinic, a sympathetic scientist, and other environmentalist supporters would participate—talks, in short, like the ones that had led to Concerned Citizens of Norco's (CCN's) relocation agreement with Shell. Orion and government officials, in contrast, imagined discussions modeled on Community-Industry Panels established by the LDEQ and Community Advisory Panels (CAPs) popular throughout the chemical industry: discussions that gave industry and community representatives a chance to get to know one another as people, that transcended divisive and polemical rhetoric in part by excluding "outside" environmentalists, and that offered the opportunity for industry representatives to better understand community concerns and clear up misunderstandings about facility operations.

Neither model, of course, fully embodied the ideal of deliberation. In fact, both activists' vision for "deliberative negotiation"[4] and industry's and regulators' imagined "dialogue" can be considered neoliberal remakings of the liberal ideal, in that deliberations addressed not the state but private companies as key decision makers on public issues. However, unlike deliberative negotiation, dialogue is increasingly common as part of a larger suite of neoliberal approaches to environmental protection and environmental justice,[5] including environmental agencies' increasing interest in framing regulated industry as partners in problem solving[6] and petrochemical companies' Corporate Social Responsibility programs.[7]

Aside from their different status within established neoliberal practices, the two models for talks diverged in their handling of issues identified by liberal democratic theorists as fundamental dilemmas for the deliberative ideal, specifically: how to cope with real power differentials between participants envisioned as equals, and what constitutes a legitimate issue for public discussion.[8] Activists' model explicitly thematized differences in power between residents and industry representatives in order to enable effective discussion. In contrast, industry- and agency-sponsored panels presumed the equality of participants, papering over differences in the interests of sociability. The two approaches to creating egalitarian discussion, notably, each had consequences for the way the scope of legitimate issues for discussion was defined. In the talks activists initiated, they were able to assert technical questions of environmental quality, community health, and plant performance as topics that warranted discussion—even if they were not able, ultimately, to gain ground in discussions of these areas. But in community-industry panels, the presumption of equality among participants gave experts opportunities to advance their understandings of these technical issues in ways that neither disrupted the reasoned, civil, egalitarian tone of discussions nor allowed for meaningful dissent. In neoliberal dialogues, then, expert knowledge of environmental health and safety became information to be communicated but not a subject to be discussed.

Even before the end of CCNS's campaign, dialogue had prevailed in New Sarpy. In early November 2002, CCNS leaders met with top Orion officials, absent Rolfes and their lawyers, to talk about terms for a possible settlement agreement; this meeting paved the way for the group's December 18 decision to drop their lawsuits. Subsequently, discussions between residents and industry in New Sarpy—as in Norco and other fenceline communities around the country—revolved around industry-sponsored Community Advisory Panels, as well as the LDEQ's Norco-New Sarpy Community-Industry Panel. These dialogues satisfied community members in that they offered the opportunity to interact with powerful refinery decision makers as equals; indeed, residents marveled at how approachable and down-to-earth industry engineers were, once you got to know them. Yet experts' very willingness to be rank-and-file participants in egalitarian discussions bolstered their authority over technical issues: by offering their knowledge as a helpful contribution to the discussion rather than pronouncing the facts from on high, they made their understandings of potentially controversial issues easy to accept, and very hard for residents to dispute without violating the spirit of discussions. A neoliberal refashioning of the deliberative ideal, community-industry dialogue thus served as a primary forum for industry scientists and engineers to assert their expert understandings of environmental and health issues while avoiding environmental justice criticisms of their claims—and to reestablish their authority over those issues in the wake of challenges presented during local campaigns.

Deliberative Negotiations

When they invited Clark Johnson to discussions, CCNS had in mind the kind of talks that had taken place between the Diamond community and Shell Chemical in Norco. From March until June 2002, at least two key Shell Norco managers, half a dozen representatives of Concerned Citizens of Norco, and several of CCN's professional allies met regularly to discuss relocating two streets in Diamond. Run by professional facilitators with backgrounds in social justice and social work, the talks eventually resulted in the Diamond Options Program, an agreement by Shell Chemical to purchase residents' homes for a fair price if residents wanted to leave the neighborhood, or, if they preferred to stay, to make interest-free loans for home improvements that would be forgiven over five years. Although the program was presented as a "victory for collaboration" in a statement issued jointly by CCN and Shell, it was widely understood by environmental justice groups and communities as a triumph for Diamond: Shell had yielded to the African American community's demands for relocation.

The process by which CCN's campaign against Shell was settled—and the process that CCNS wished to emulate in their dealings with Orion—appears at first glance anything but deliberative. Negotiations are considered a form of pluralist politics, in which groups with predetermined interests compete to get the best deal for themselves: CCN fights for relocation, for example, while Shell Chemical tries to get the community to concede that it is a good neighbor. With participants acting out of self-interest, the outcomes of pluralist processes tend to favor wealthy, powerful groups over those of less advantaged citizens—Diamond residents' victory notwithstanding. Deliberation, in contrast, rules out narrow self-interest. Generally seen as the major alternative to, and antidote for, pluralist approaches to democratic politics, deliberative processes ask citizens to set aside their individual or group interests and focus on the common good, giving and listening to reasons why one course of action or another would be beneficial for the entire polity.[9] Diamond residents' single-minded pursuit of relocation in their talks with Shell would seem to violate this central norm—making the deliberative overtones of CCNS's calls for talks mere window dressing.

Yet when participants in the CCN-Shell talks describe how they arrived at the Diamond Options Program, they describe not a struggle for advantage but a process of exchanging reasons and arriving at an improved, mutual understanding of their common interests. They describe a deliberation. What their accounts make clear, though, was that the deliberative quality of the actual talks depended on a series of other activities not usually associated with deliberation: candlelight vigils, accusatory websites, and behind-the-scenes conversations among powerful people, a few of whom were sympathetic to Diamond residents' cause. The case can be seen as an exemplar for community-industry talks in places like New Sarpy, not because Diamond residents "won," but because it speaks to the conditions necessary to enable community members and industry officials to engage in reasoned discussion as equals, including discussions of technical topics. Specific interests did not need to be checked at the door, it turned out—but critical engagement with the power inequalities between participants was crucial.

Reasoned Understandings

When Wayne Pearce, Shell Chemical Norco's plant manager at the time of the 2002 negotiations, talks about how the company came to propose the Diamond Options Program, he describes a process with many of the hallmarks of deliberation. In September 2000, Shell had offered to buy the homes of residents on the two streets nearest to its west site fenceline—half of the Diamond community. Community protests began almost immediately: CCN charged the company with splitting up families and fracturing community networks on which elderly residents depended. But it was not until the negotiations, according to Pearce, a forty-something white man from Wales, that he and his colleagues at Shell understood the impact of the partial buyout:

> We didn't set out to break people apart, we thought we were doing something nice. It didn't turn out to be something nice in reality, for some of the people, the caregivers were split. . . . At first, we'd say, "But it was voluntary. We didn't take bulldozers in there and move people out. We offered people money for their homes. And if it was splitting up a family, well, they didn't have to take the money." But you go and sit and you realize, of course they had to take the money. It was an opportunity in their life they'd never ever seen. . . . I was born in a small south Wales coal mining village. My father was a coal miner. . . . And that made me think about that a little bit. The

premium payment on the properties [is] probably more than some people are going to earn in the next three years. How is that, is that a real choice?

Pearce's account suggests a deliberative process in several respects. It speaks, first, to the exchange of reasons that deliberation demands: Diamond residents had been advancing Shell's partial buyout, and its effects on social networks, as a reason Shell ought to relocate the remainder of the community. Through the negotiations, Pearce and other Shell officials finally came to understand it *as* a reason: looking at the partial buyout from the perspective of residents, they realized that, as "an opportunity in their life they'd never ever seen," residents on the two streets nearest the plant did not really have a choice to stay to preserve their community networks.

Other reasons for relocation were given during the negotiations as well; in keeping with the promise of deliberation, these helped persuade Shell decision makers. Through the talks, according to accounts by other participants, Pearce came to understand Shell's history of racism in Norco and its consequences for the community. Iris Carter, a fifty-year-old black woman who represented Diamond residents in the talks, recalled describing how Shell excluded the black part of Norco in the 1950s and 1960s:

I was sitting by Wayne one day, and I remember, I was telling a story about how they used to do us when we were younger, they wouldn't let us come to the Plant Day, they had bowling alley, movies, girl, we couldn't go to that. Couldn't come to that. They didn't give you no job, anyway. So I could hear Wayne say, I could hear him say, "What did we do to these people?"

That moment of realization by the plant manager, as Carter remembered it, was pivotal in the negotiations. Steve Lerner's account of the campaign corroborates the idea that understanding racism as a reason for relocation helped lead to Shell officials' creating the Diamond Options Program.[10] According to Lerner, Peter Warshall, an influential environmentalist with connections to the CEO of Royal Dutch Shell who sat in on some of the negotiating sessions,[11] explained to Pearce Louisiana's long history of racism and corrupt politics and argued that residents' demand for relocation needed to be understood in that context. In talking to Lerner, Warshall suggested that his intervention had helped give Shell a reason to relocate residents: the injustices that they had previously suffered, in part at Shell's hands.

Besides pointing out the exchange of reasons that occurred as part of the negotiations, Pearce's own account of how he was persuaded by Diamond residents evokes two other elements of deliberation. He describes relating to the economic situation of Diamond residents through his own experience; there would have been no "choice" about an opportunity like the partial buy-out where he grew up, either. In making that connection, Pearce imagined a fundamental commonality between himself and residents, a commonality that undergirds deliberative theorists' expectations that people can come to agreement through reasoned discussion. Beyond their shared humanity, Pearce and Diamond residents were subject to, and motivated by, a common economic system.[12] The shared system made residents' reasons intelligible to Pearce, even though he occupied a different place in it.

Finally, when he talks about doing "something nice" for the community, Pearce also alludes to a common good—an idea of what is good or bad for the community, on which there can be agreement. Fundamental to deliberation, the notion appears as pivotal in other participants' accounts, as well. Iris Carter described to me how, in her view, the negotiations changed Shell's approach to CCN:

> I got to know them as people and they got to know us as people, because when you, when it started out, I think we were just statistics on a piece of paper to them. . . . But once they saw that we were human beings, we were intelligent, we weren't trying to get over, we weren't just trying to hit them up for money. We were hitting them up for a better life that we were entitled to! And I think that changed their hearts. . . . Once they got to understand where we were coming from and who we were, that made a difference in their choice to do better, to try to make it better for us.

In addition to evoking the fundamental commonalities that make deliberation possible, Carter's account emphasizes the important role that a notion of the common good played in furthering the discussions. In contrast to "hitting them up for money"—the kind of narrow self-interest that deliberation disallows—the goal of "a better life" was something that, in Carter's assessment, Shell decision makers were willing to go along with.

Pearce and Carter's descriptions suggest that, although the talks between Shell and Diamond were billed as "negotiations," they had important deliberative traits: residents and Shell officials gave each other reasons, to persuade rather than compel certain courses of action. Their reasons referred to the long-term good of the community, rather than the narrow interest of anyone in particular. And they came to appreciate one another's reasons by finding areas of commonality as human beings embedded in a shared economic system. In that sense, these interactions satisfied the deliberative norm—reasonable discussion among equals of the common good—invoked by calls for "talks" in New Sarpy a few months later.

Empowering Deliberation

Accounts that highlight the deliberative aspects of negotiation emphasize the commonalities between Diamond residents and Shell managers: they have the potential to identify a common good; they can identify with one another as people; they are equally able to advance reasons for their favored courses of action and to persuade others with their reasons. Focused narrowly on the negotiations themselves, such accounts thus invite us to ignore the substantial inequalities between low-income, African American residents of Diamond and well-paid, highly educated representatives of the Shell Chemical company. The power differentials come into focus only when the negotiations are viewed in the context of CCN's larger campaign, which started many years prior to the talks and continued until the community group finally reached a settlement with Shell. Activities associated with the campaign aimed at "empowering" residents with respect to the company—giving them the clout to influence Shell decision makers. These activities, it turns out, also made the deliberative aspects of the talks possible.

Rallies, marches, press conferences, and similar tactics used by CCN and other community groups to challenge petrochemical companies are usually considered orthogonal to deliberative processes: democratic theorists contrast direct action and reasoned deliberation in discussions of how citizens can best participate in democracy.[13] Yet environmental justice professionals do not spurn deliberation; in fact, talks between community groups and plant managers are seen as an essential part of a winning campaign. In his *Good Neighbor Handbook: How to Win!* veteran community organizer Paul Ryder, the white, middle-aged organizing director at Ohio Citizen Action, an environmental justice nonprofit headquartered in Cleveland, encourages communities to talk to the companies that they oppose.[14] Comprising an entire chapter of the book ("Talking with the Company"), his advice even echoes deliberative ideals in places. For example, Ryder counsels communities not to go into talks with "demands"—a concept not unlike that of "narrow self-interest" in deliberative theory—but to identify their underlying goals and use them as the basis for finding common ground with industrial facilities.

However, unlike deliberative approaches that presume commonality or equality among residents and plant officials, Ryder's and other professional activists' approach to talks focuses implicitly on the *in*equalities between the two groups. Ryder's advice stresses the need for community groups engaged in campaigns against local facilities to take and keep "the initiative"—the ability to define the campaign and its issues. When a community has the initiative, Ryder explains, the company can only react to the community's activities. Community members, in short, have power in their dealings with the company; it can no longer afford to simply ignore them.

Talks are an outcome of this community empowerment: "as the campaign escalates," he predicts, "the company eventually realizes it is in their interest for a real decision-maker to talk with neighbors in earnest."[15] Yet the initiative—the community's increased power with respect to the company—must be maintained for the talks to be successful:

> It is essential that the campaign accelerate during the negotiations. This shows that you know the campaign is the only reason why talks are being held. The stronger the campaign, the sooner a good outcome will emerge.
>
> For the same reason, if the campaign goes into limbo once talks start, you can bet that the talks will also go into limbo before long. You have just taken the initiative from yourself.[16]

In Ryder's analysis, then, talks like those in Norco—talks in which community members and industry representatives are able to achieve a sense of commonality through reasoned discussion—occur only as a result of community groups' work to lessen the power imbalance between residents and company officials.

Viewed from Anne Rolfes's perspective, the settlement in Norco unequivocally supported Ryder's advice.[17] Speaking to me just weeks before New Sarpy residents ended their campaign, Rolfes explained the relationship between talking and campaigning in terms very similar to those used by Ryder:

AR: Well, they wouldn't meet with us. Shell wouldn't meet with us, neither would Orion. So you beat up on them until they will.

GO: And once they do?

AR: Well, once they do, we still need to keep active. When we were in negotiations with Shell, we said, just because we're talking to you doesn't mean that the campaign stops. The campaign stops when we have resolution. It doesn't stop just because we're sitting down. Obviously. If it stopped, they would lose their incentive to deal with you. So it ceases once the community is satisfied with the resolution.

To underscore the importance of continued campaigning—of, that is, ongoing activity to force industrial facilities to respond to the community—Rolfes described an event held by CCN midway through their negotiations with Shell. After months of talking, Diamond residents and their allies came to believe that Shell Norco officials were not serious about reaching a relocation agreement. To increase the pressure on them, CCN, Rolfes, and other supporters demonstrated in front of Shell's American headquarters in Houston. Pearce and the other local representatives involved in the negotiations were "livid," according to Rolfes; however, with their bosses even more eager to see the situation resolved, they could not abandon the negotiations. On the contrary, Rolfes believed that the event was crucial in moving them closer to a relocation agreement.[18]

In their recipes for talks, environmental justice activists' attention to power imbalances between industrial facilities and community groups extended beyond their focus on campaign activities as a way to seize "the initiative." Activists like Rolfes were also concerned that talks themselves be structured in such a way that residents would not be overpowered by industry representatives. Together with CCN, they saw to it that talks not be limited to residents and Shell Norco decision makers. Like many of its industrial peers, Shell would have preferred their interactions with residents to be unmediated by "outsiders." Activists, however, demanded that the environmental professionals who had been most involved in the campaign also be participants in the negotiations. As a result, not only Rolfes, CCN's chief advisor on everyday matters of strategy, but also CCN's attorney Monique Hardin, an African American environmental lawyer with the nonprofit EarthJustice, and Wilma Subra, a white chemist known for her work in support of Louisiana's communities, were among those on hand to support Diamond representatives during their discussions with Shell.

CCN and their allies also insisted that the talks be run by professional facilitators—by "neutral, third-party moderators," in the language of New Sarpy residents' newspaper "invitation" to Orion's CEO. Although Shell agreed readily to this condition in principle, in Rolfes's recounting, the company and the community struggled over who exactly could be considered a "neutral" third party. Among the candidates proposed by Shell was a male professor from a New Orleans university whose approach to community-industry conflict, while much admired by industry, was thought by activists to be biased in favor of companies. After significant conflict on the issue, participants finally agreed on two female facilitators with backgrounds in social work and social justice.

Where campaign activities worked to increase residents' power in their interactions with industry, the talks' structure aimed at ensuring that the balance of power achieved by residents through their campaign carried over to the talks themselves. Speaking to me about the facilitators after the negotiations had concluded, Anne Rolfes made clear her understanding of their purpose: to ensure that *community members* were heard. To her, the facilitators had often seemed too concerned with how Shell representatives felt, asking them if they found LABB and Concerned Citizens of Norco's (CCN's) activities objectionable even before they had said anything—when, in Rolfes's opinion, *they* were perfectly capable of speaking up if they had a problem.

Despite the deliberative ideal invoked in activists' calls for talks, then, and despite participants' subsequent descriptions of the Norco negotiations' deliberative elements, talks in Norco involved—and depended on—explicit engagement with the inequalities of power between Diamond residents and Shell Chemical. Through campaign activities, community members were able to force decision makers at the powerful company to negotiate with them. Moreover, having taken "the initiative," they were able to ensure that power imbalances would be addressed even in the talks themselves—through the support of more powerful allies and the intervention of facilitators attuned to justice issues.

Asserting Expertise

An exemplar of environmental justice activists' model for community-industry talks, the negotiations in Norco combined deliberative content—reasoned discussion with attention to a common good—with an explicit negotiation of power. Rather than assuming that Diamond residents and Shell officials could discuss issues as equals because of their common status as citizens and human beings, activists from CCN and other organizations worked at creating greater equality among participants in negotiations through activities and structures that reduced Shell's ability to steamroll or simply ignore residents.

Yet power differentials remained—and, in the negotiations in Norco, were manifest especially in the conflict over health. Facing Diamond residents' claims that Shell's emissions made them sick, facility decision makers exerted their power as representatives of a multinational corporation to declare the topics off-limits. As Diamond representative Iris Carter told me, "They told us if we talked about health issues, my people from Norco, from Concerned Citizens, told us if we brought up health issues, they wouldn't gonna negotiate to get us out of here." By reminding residents that they could simply continue operating without dealing with the community, Shell representatives invoked their company's status—large, wealthy, and subject to control only through very specific (and limited) kinds of government intervention. Importantly, their refusal to talk about health was not just an expression of corporate power but also an assertion of scientific authority.

According to David Brignac, Shell's white, thirty-something Good Neighbor Initiative manager, he and Pearce refused to discuss health issues in the negotiations because Shell experts were certain that their operations did not affect residents' health. The company had done studies, Brignac told me in a December 2002 interview, that showed that workers at its Norco plant were, in fact, healthier than the U.S. average—leading them to conclude that any higher rates of ill health in Louisiana's industrial corridor had to do with access to health care and other "lifestyle" factors rather than chemical exposures. This evidence, Brignac recalled, was distributed during the negotiations as justification for the company's refusal to discuss health:

> We gave copies of the study to the Diamond residents we were negotiating with, and to Anne [Rolfes], Monique [Hardin], Wilma [Subra]. So that did come up, and we said, "Look, we've got this one study. We've done more studies, but this is the most recent one. And this is what it shows." And we gave them the report. We didn't really discuss it at length or anything, but we said, "Our position is, we're not harming health, and so, as a result, we don't want health to be a part of this negotiation. In other words, if you're trying to negotiate with us on the basis we're harming your health, we're not going to negotiate. So you're going to have to find some other reason for wanting a buyout, because we're not going to agree that it's because we're harming your health."

In the position taken by Shell, power and expertise were thoroughly intertwined. The company's ability to produce scientific studies was offered as a reason that health was not a legitimate topic for negotiation. At the same time, Shell's power to walk away from the negotiations prevented challenges to those studies by residents whose experiences living near Shell led them to make conflicting scientific claims.

It is not surprising that corporate power and scientific expertise would reinforce one another. What is notable here is that Shell's claims to technical expertise could only be made authoritative in conjunction with an overt show of force. Brignac and Pearce, both engineers by training, had to assert Shell's ability not to engage with the community in order to make studies conducted by their scientist colleagues have weight in the negotiations. CCN and its allies acquiesced to the prohibition on talking about health—and, effectively, to Shell's claims about health effects—because they saw the possibility of negotiating a settlement on other grounds. As we shall see, other forms of talks, in which inequalities between community and industry were not made explicit, made such crude reminders of industry engineers' and scientists' power as corporate representatives unnecessary.

Egalitarian Dialogue

According to Jason Carter, one of Orion managers' first priorities upon taking over the refinery in 2000 was to establish an "active dialogue" with the community. Initially, they were frustrated. The first meeting with residents that the company set up in September 2000 was, in Carter's words, "an hour of accusations and hostility from the community and the [Louisiana] Bucket Brigade." Subsequently, Orion was unable to get community leaders to "just sit down and talk," especially absent LABB. The following year, still seeking dialogue, the company set up its own Community Advisory Panel (CAP), following a model used widely in the chemical and petrochemical industries and advocated by the American Chemistry Council, the industry's major trade group, as part of their Responsible Care program.

Just as activist-initiated negotiations are often assumed to serve the narrow self-interests of residents, industry-sponsored Community Advisory Panels are frequently dismissed as shallow public relations stunts. Critics complain that they avoid substantive engagement on environmental issues by excluding the most critical voices from the community.[19] Indeed, while Orion's CAP initially included two of CCNS's officers, its conveners purposefully diluted opposition by drawing members not just from New Sarpy but from all over St. Charles Parish.[20] In the context of the petrochemical industry's larger push to avoid or limit government regulation—and regulatory agencies' drive to devolve problem solving to the local level—CAPs and their cousins, LDEQ-sponsored Community-Industry Panels, appear to manufacture consent under the guise of "dialogue."[21]

Yet, as with negotiations, CAPs evoke key deliberative ideals not only in their rhetoric of "dialogue" but also in the way they organize discussion among residents and industry representatives. Specifically, beyond paying lip service to "two-way communication," the CAPs and Community-Industry Panel in St. Charles Parish were organized to promote an egalitarian atmosphere and to focus discussion on the common good rather than on particular interests. These features ultimately gave CAPs credibility in the eyes of residents: by the spring of 2003, several more CCNS leaders had joined

Orion's CAP, and by 2006, the New Sarpy residents who had been the most active members of the opposition to Orion valued the CAP established by Valero for the relationships it allowed them to build with the refinery's new managers.

Nonetheless, community-industry dialogues *did* limit dissent—and not through some failure of commitment to their deliberative ideals. Rather, the very ways in which the CAPs and Community-Industry Panel put those ideals into practice—the way they structured egalitarianism and bounded discussion of the common good—restricted resident panelists' ability to raise the most controversial issues and, especially, stifled challenges of industry scientists and engineers' claims about pollution, plant safety, and community health. Obfuscating power differences, the CAPs' form of deliberation thus enabled experts to exert their authority over technical issues without explicit shows of force that would disrupt the mutually respectful tone of dialogue with residents.

Structuring Equality

At the Valero CAP's February 2006 meeting, CAP members worked together to pick topics for their bimonthly meetings in the upcoming year. An annual ritual conducted similarly in the St. Charles CAP and the Norco-New Sarpy Community-Industry Panel, the agenda-setting exercise helps exemplify how egalitarianism was structured on the panels—namely, by assigning identical roles to all CAP members, regardless of their expertise or social status. At the meeting, CAP members representing Valero and members representing the community, mostly residents of New Sarpy, sat interspersed at a U-shaped table. Having finished their meal and heard announcements, members were directed by the CAP's facilitator to brainstorm, in pairs, areas of common interest and concern between the community and the refinery that would be appropriate topics for CAP meetings. The various pairs' ideas were then listed on a flip chart at the front of the room, and each member was instructed to vote for one top choice. The four topics with the most votes were then slated for the upcoming meetings.

The structure of the agenda-setting process thus made everyone the same: suggestions by Valero representatives were not given special status; their votes did not count more than those of representatives from the community. This structural sameness carried through the activities of all of the community-industry panels on which New Sarpy residents served. On each, representatives of industry and representatives of the community were equally "members." CAP meetings were in general attended only by members—though sometimes residents potentially interested in join-ing the CAP would attend as the guest of a member. Community-Industry Panel meetings, in contrast, distinguished between formal members of the panel, who sat together at tables along one side of the room, and the inter-ested public, who occupied rows of chairs in the center. But the distinction applied to representatives of industry as well as to residents: when facility managers who were not formal members of the panel attended, they sat in the audience.

The structure of panel programs themselves also avoided setting indus-try members apart from resident members. The CAPs and the Community-Industry Panel all included announcements at the beginning of their meet-ings. While the announcements gave industry members an opportunity to share information about their plants' accidents, achievements, and finan-cial status, they offered resident members a parallel opportunity to report on events in the community. During announcements at the February 2006 Valero meeting, for example, industry members introduced the refinery's new maintenance and operations manager, announced that the plant had been honored within the company for its safety record, and reported on the progress of a major beneficial environmental project they were undertak-ing as part of Orion's 2002 settlement with the LDEQ. Resident members, in turn, announced the formation of New Sarpy's first Girl Scout troop, briefed the group on the effects of the parish's post–Hurricane Katrina population surge on local government services, and encouraged fellow residents to take advantage of educational programs at the Council on Aging. Differences in content notwithstanding, the parallel sets of announcements gave industry and resident members symmetrical opportunities to speak with authority on events of common interest.

Industry and resident members similarly played identical roles with respect to the programs that were the centerpiece of CAP and Community-Industry Panel meetings. Devoted to themes such as "Emergency Prepared-ness" or "Environmental Risk," most meetings featured presentations by one to three individuals knowledgeable about the theme. However, despite their own considerable knowledge, industry members almost never played the role of presenter. Instead, government officials, research scientists, local busi-nesspeople, and the facilities' own midlevel specialists were invited as speak-ers. Members from industry and members from the community together comprised the audience, listening to the presentations and asking questions of the presenters. Soliciting presenters from outside the CAP downplayed the disparity in expertise between industry and resident members and rein-forced the structural equality among panelists.

The equal status afforded to resident and industry members made possible the kind of mutual recognition on which deliberative processes depend. Playing a common role as panel "members," representatives of facilities and representatives of the community shared meals, engaged in small talk, and came to learn about one another's families and extramural interests. Drawing on their basic commonalities—as parents, as homeowners, as people with choices to make about how to take care of their families—some resident CAP members sought to build relationships with their industry counterparts: after one Orion CAP meeting, Don Winston and Jason Carter tried to find a date when they and their significant others could all go out to dinner. Resident members also situated their understanding of environmental issues at facilities with reference to their understanding of industry CAP members as ethical human beings. Audrey Taylor, for example, told me that she believed that the people who ran the nearby plants were doing the best they could to be safe and comply with the laws; despite her concerns about their effects on community health, she trusted that the industry representatives she had gotten to know on the CAP were interested in protecting the community.

The formal equality of CAP and Community-Industry Panel members did not, of course, eliminate or render irrelevant the differences in their power and expertise. As we will see below, industry members exercised a great deal of control over the messages conveyed in panel presentations—without violating the structures of egalitarianism. Inequalities also colored the sociability and trust fostered by the panels. Resident members were very aware that their rapport with industry members was forged across important differences. In speaking to me about their participation on the CAPs and Community-Industry Panel, residents marveled at how approachable the representatives from industry were; how "down-to-earth" they seemed; and how patient they were with explanations. But far from threatening residents' sense of egalitarianism on the CAPs, acknowledging the disparities of power and expertise made egalitarian dialogue seem that much more of an accomplishment to residents—as, in contrast to their implicit expectations, they found they *could* interact with industry members as equals.

By giving all panel members the same formal status, Community Advisory and Community-Industry Panels paved the way for industry and community representatives to recognize one another as human beings with shared interests and experiences beyond their training or professional affiliation—and in turn created the possibility of reasoned deliberation. However, in asserting structurally the equality of members from industry and the community, the panels also foreclosed the possibility of acknowledging *inequal-*ity in any overt way. During meetings, resident members often thanked or praised industry members for some aspect of their interaction with the community, from funding a community project to listening to community concerns. However, that they did not need to do so was left implicit, and resident members did not marvel publicly at how industry members' behavior defied their expectations of how powerful experts would act. Residents' arguably inferior status was even less legitimate as a subject for discussion. While CAP and Community-Industry Panel programs sometimes focused on jobs or economic development, community need was framed in terms of education rather than empowerment. Moreover, resident members of the panels were assumed not to be among those in need, but among those positioned to help educate neighbors trying to better themselves.

A basic deliberative value, egalitarianism was thus implemented quite differently in community-industry "dialogue," as represented by the Community Advisory and Community-Industry Panels, than in activist-initiated "negotiations." Negotiations sought equality through attempts, within and outside the discussions themselves, to increase residents' power with respect to industry representatives. Dialogue, in contrast, asserted equality through structures and practices that made resident and industry members all the same—or, rather, that ignored their differences. Further, where the mutual recognition achieved in negotiations depended on building up residents, that produced in dialogue rested on industry members consenting to come down to residents' level, by making themselves sociable people with homes and families.

Bounding Discussion

Deliberative ideals were evidenced not only in the egalitarian form of community-industry dialogues but also in the substance of discussions. In keeping with the requirement that deliberation transcend particular interests and focus on the general good, panel presentations and the questioning that followed framed issues—issues that often had immediate, personal consequences for CAP members—in a general, indirect way that downplayed both industry and community members' stakes in the issue.

For example, dedicated to the topic of job opportunities at chemical plants, the St. Charles CAP's May 2003 meeting featured three presentations: about a vocational program at the parish high schools; about the Louisiana Technical College's (LTC's) associate's degree program for would-be chemical plant operators; and about how applicants for jobs at the plants are recruited, screened, and ultimately hired. Although three of St. Charles Parish's five largest chemical facilities, as well as several smaller ones, were represented on the CAP, the presentations made scarce mention of any particular company's hiring policies or job training efforts. While the one company employee who presented (the other two speakers represented the LTC and a consulting company) acknowledged his and other companies' participation in the high school training program, the focus of his presentation was the opportunities that the program offered to any student who wanted to work in the chemical plants.

The jobs program's emphasis on the hiring process and vocational training programs—matters presumably of interest to anyone who lived or worked in the parish—distinguished it from mere public relations for the companies represented on the CAPs. The questions and comments that followed the presentations likewise framed employment at the plants as a general issue. One white, male resident member asked what the average age of a new hire at one of the plants was—he said that he wanted to be able to reinforce to his high school science students that they would have to work toward their goals for a number of years after graduation if they were to achieve them. Mitchell Mobley, speaking as an industry member, added to the comments of the speakers by stressing that companies now hired residents on the strength of their skills and qualifications, and not on the basis of family or political connections as was the practice in the past. Notably, these and other questions focused on high-level issues rather than the situations of any plant workers, families, or communities in particular. No resident member asked, for example, "My nephew had a straight-A average at Louisiana Technical College. Why did Shell not hire him?" And no industry member talked about his company's increasing preference for hiring contractors or its competitive benefits package. No one even asked the industry representatives how they liked their jobs or how they got hired at a chemical plant.

Indeed, raising questions that made explicit members' personal stakes in the issues under discussion was taboo in the context of CAP and Community-Industry Panel discussions, and residents trying to do so were made to feel unwelcome. In August 2002, at the height of the campaign in New Sarpy, Orion CAP member and CCNS leader Don Winston went to a CAP meeting prepared to present the community group's demands to Orion officials and confident that other resident members would be supportive. He returned disgusted: he told me that he had gotten "beaten up" so badly that he was ready to resign his position on the CAP. It is certainly possible that other members were not sympathetic to the substance of CCNS's demands—nonmembers were banned from Orion CAP meetings during that contentious period, so I did not witness the interaction. However, given Winston's initial assessment of their positions and my subsequent observations of the general-good focus that the Orion CAP, like the St. Charles CAP, adopted, it seems likely that Winston's fellow resident members were angered less by the demands themselves than by the fact that he violated the norms of discussion by making demands at all. The campaign demands of Diamond residents were similarly unwelcome at Norco-New Sarpy Community-Industry Panel meetings: Anne Rolfes reported that at one meeting that CCN members attended not as panelists but as members of the public, their comments about Shell's treatment of the African American community were shunted to the very end of the meeting, and a time limit was imposed on speakers. Again, while the substance of their demands may have been objectionable, the very making of demands was disruptive to the structure of dialogue and had to be confined to the margins of the meeting.

On the rare occasions when resident members successfully raised particular community issues with specific chemical plants in the context of panel discussions, the issues were reframed in terms of the general good. At the December 2002 Community-Industry Panel meeting, the facilitator's call for "the word on the street"—the middle-aged white woman's tellingly generalized way of asking for announcements from resident members of the panel— yielded a pointed comment from Margie Richard, the sixty-something black woman who had spearheaded Diamond's campaign for relocation. "People are wondering," Richard said, what had become of Shell's promise to establish a memorial at the site of the Bethune School, the black high school that burned down on the eve of educational integration. Although Richard's comment singled Shell out, it was cast in terms of a general good: it was not a demand for action but a request for information; moreover, it was not Richard or CCN but "people" who wanted to know.

Distancing themselves from overt statements of specific interests by either resident or industry members, CAPs and Community-Industry Panel discussions implemented a central ideal of deliberation—and distinguished themselves from mere industry public relations. At the same time, they effectively excluded from discussion whole domains of community concern by presuming that community members' (and industry employees') individual experiences of living near, and with, petrochemical facilities were not relevant to the general good. Such a presumption is not necessary to deliberative processes, as negotiations between Shell and CCN demonstrate. In that case, Diamond residents—empowered through their activism and the structure of talks—were able to convince Shell decision makers that, in relating their experiences of racism at the hands of the chemical plant, including their historical inability to get jobs there, they were presenting not self-interested gripes but information important to understanding how to secure a better life for the community. Through their discussions, the particulars of their situation were translated into a matter of the general good. This translation—which deliberative theorists argue is essential to including historically marginalized groups in deliberative democracy[22]—is, in contrast, made impossible by the manner in which community-industry dialogues implement the same deliberative norms. With personal experiences and stakes in issues ruled out of bounds from the start of the discussion, resident members of Community Advisory and Community-Industry Panels had no opportunity to expand ideas of what might constitute an issue of general interest. In combination with the panels' structural assertion of equality, presumptions about what was and was not appropriate for discussion helped shield industry members' expertise from critique—without overt assertions of scientific authority.

Speaking for Science

Although a few CAP and Community-Industry Panel programs focused on jobs or economic development, the majority of them were devoted to technical topics. Indeed, one of the explicit goals of Community Advisory Panels is educating community members about plant operations and performance.[23] At least two Community-Industry Panel meetings that I attended, for example, included discussions of Shell's and Motiva's air monitoring program in Norco; an Orion CAP meeting featured a presentation about the parish's water treatment system, located just downriver from Orion and recently upgraded with the refinery's help. Minutes of St. Charles CAP meetings going back to the panel's establishment in 1992 show programs on emergency preparedness, emissions, and environmental health to be perennial favorites.[24]

Each of these topics was potentially controversial: in other settings, CCNS members questioned Orion's handling of a thirteen-hour fire in a multi-million-gallon gasoline storage tank; bucket users in both New Sarpy and Norco challenged plants' accounts of their air emissions; and many residents of the area were convinced, despite industry's reassurances, that their health was compromised by chemical emissions. Yet in the CAPs and Community-Advisory Panel, these issues did not become contentious—in large part as a result of the way the panels implemented deliberative ideals.

As equal "members" of the panels, industry representatives were able to exert disproportionate influence over the framing of technical issues without disrupting the meetings' egalitarian tone. Although industry members officially had no greater say than resident members in determining CAP agendas, their contributions to agenda-setting discussions steered the panels to approach topics in ways accepted by industry experts. For example, the minutes of a 1992 St. Charles CAP agenda-setting meeting record the following interaction:

> It was suggested that the panel consider the issues of pollution and environmental health.
>
> Panel members were then asked about their specific concerns regarding the environment. In response, Panel members raised such issues as emissions and the impact they have had on the environment. They wanted to know more about the incidence of diseases from specific environmental carcinogens, the symptoms of such diseases, and what can be done to lower the percentage of risk associated with emissions? They also wanted to hear the whole story: what were the conditions 10 or 15 years ago, how do those conditions compare with today, and where are we going in the next 10 to 15 years.
>
> One plant representative cautioned that the proposed topic was very broad and recommended that the Panel start with the risk and health topics and then move to the issue of emissions. This suggestion was accepted by Panel members.

The broad issues identified by resident members—ranging from risk to general environmental conditions and impacts—could have been the basis for a program that raised open-ended questions without clear-cut answers; for example, what effect *does* industrial pollution have on air and water quality? But the possibility was forestalled by the unnamed industry member's contribution. Without stepping outside his role as a rank-and-file CAP member, he reframed residents' concerns to be consistent with the way industry experts understand the issues of environmental performance; that is, in terms of health, emissions, and risk. Other members could have rejected his suggestion but did not, no doubt finding it, as he intended, a useful way of breaking down a massive topic.

By assenting to the industry member's framing of the issues, CAP members also accepted limits on the questions they would be able to raise in subsequent discussions. A meeting focusing on "pollution" might consider overall measures of environmental quality in St. Charles Parish and ways the petrochemical industry there has contributed to poor water quality in the bayous and river. In contrast, CAP meetings about "emissions," which occurred relatively frequently, discussed what quantities of which chemicals were being released by individual facilities. These discussions divorced quantities of emissions from questions about environmental quality, comparing current emissions data only to data from prior years—and demonstrating that emissions were being progressively reduced. Conversations about "risk," similarly, tend to replace questions about, for example, community well-being, with probabilistic assessments of disease rates and other narrowly defined outcomes.[25]

In addition to suggesting how topics should be framed, industry members influenced the shape of technical discussions by suggesting and recruiting speakers for programs. Once a CAP or the Community-Industry Panel had decided on a list of topics, the group's facilitator would ask members for their help in finding presenters for each program. Although resident members were, in theory, just as welcome to nominate speakers as industry members were—and, indeed, they suggested local businesspeople for economic development programs—it was industry members who volunteered to invite their professional colleagues to speak at meetings. As in the case of agenda setting, their efforts were seen as helpful contributions to the group rather than as attempts to manipulate it, and resident CAP members inevitably accepted their suggestions. Once again, industry members' de facto control over speaker selection limited the possibilities for resident members to raise critical questions about industry's technical claims. In the St. Charles CAP's programs on health, for example, speakers included industrial hygienists and epidemiologists from chemical companies who surveyed the companies' studies of occupational health, and representatives from the Louisiana Tumor Registry who presented statistics that showed that cancer incidence rates in Louisiana were no higher than in other parts of the country. Researchers raising questions about Louisiana's cancer statistics or pioneering methods for studying community health were not represented among the CAP speakers recruited by industry members.

Although industry members' participation arguably shaped CAP deliberations in a way that privileged expert understandings of environmental and health issues and avoided lines of questioning that would be threatening to the industry, it did not disrupt the egalitarian practices that structured CAP meetings. On the contrary, the structural equality of industry members and resident members helped enable industry representatives to steer the discussion without overt exertions of their authority. Their status as members prevented them from declaring what must be discussed or prescribing how the discussion should proceed and instead framed their interventions as helpful suggestions, subject to the group's approval.

Yet the actual inequalities between industry and resident members of the panels helped ensure that industry experts' issue framings and speaker suggestions would be approved. As (in most cases) trained scientists and engineers with (in all cases) extensive experience in petrochemical plants, industry representatives were steeped in a series of frameworks for analyzing and managing the complex interactions among industrial processes, the natural environment, and human health. Whatever the failings of experts' frameworks, they offered tools for discussing issues of community concern in bite-sized chunks. For resident members to have challenged experts' framings, they would have needed to have comparable frameworks—heuristics for approaching interrelations between environment and health as residents understood them. While such frameworks do exist in the environmental justice movement, even those resident members who had participated in community campaigns (a small subset of CAP members) had no formal training and but a few years of nonprofessional experience mobilizing them, giving them relatively little capacity to formulate compelling alternatives to expert framings. Similarly, industry members' professional networks—which extended beyond their companies to regulatory agency experts, peers at other companies, and fellow alumni of engineering and science degree programs—gave them access to numerous credentialed individuals who shared the same frameworks and who could be called upon as presenters. Resident members had no such networks; their only contact with alternative experts would have been in the context of community campaigns, in interactions arranged and mediated by groups like the Louisiana Bucket Brigade. Despite these scientists' impressive credentials and affiliations with respected academic institutions, their association with LABB and environmental justice campaigns led industry to question their legitimacy—leaving resident members, again, unable to offer alternatives to industry members' suggestions.

Not only the CAPs' structures of equality but also their way of distinguishing acceptable topics for deliberation—topics pertaining to the general good—from unacceptable ones—having to do with particular interests or personal experiences—made it difficult for resident members to raise or pursue technical questions in a manner that might be challenging to industrial expertise. As meeting themes suggest, scientific topics were taken to be of general interest. However, in keeping with the prohibition on particular interests or experiences, only information that could be presented in an abstract, depersonalized manner had a place in Community Advisory and Community-Industry Panel discussions. Scientific studies, technological processes, and government regulations and procedures—all having status as entities independent of their authors' interests and identities[26]—were, accordingly, the focus of panel presentations. But questions about the facts presented also had to be cast in the same, abstract way, limiting the ability of resident members to probe the uncertainties or limitations of the science, technology, and regulatory frameworks explained by expert presenters.

Questions asked at the Norco-New Sarpy Community-Industry Panel meeting in April 2003 exemplify the ways in which community members' environmental and health concerns were transformed to meet the requirements of deliberation, at least as implemented in community-industry dialogues. In a program devoted to discussion of Shell's and Motiva's "Air Monitoring...Norco" study, Shell's Good Neighbor Initiative manager David Brignac explained how the study was designed, including how its results were to be disseminated; a white male representative from URS, the independent environmental firm that carried out the air sampling, presented the results of the first phase of monitoring—that Norco's air met state standards, was relatively uniform, and was comparable to air in "other cities"; [27] and Luann White, a white professor of public health from Tulane University, spoke about the community health survey planned as part of the study. An unusually lively discussion followed: eight different resident panelists and audience members asked questions, which were addressed not only by the speakers but also by several of their colleagues who were sitting in the audience.

In keeping with the deliberative ground rules for discussion on the panel, the questions largely excluded the personal or experiential dimensions of community members' concerns about air pollution and its effects on health. A number of the questions focused on pollution and its sources: Ram Ramachandran, an Indian American male representative to the parish council, asked how one could know how much pollution came from the plants and how much from natural sources; another, white man wanted to know whether chemical concentrations at one monitoring site in particular could be attributed to highway traffic. A few focused on experts' procedures and methodology: the Norco Civic Association's white senior citizen president, Sal Digirolamo, questioned whether Centers for Disease Control data could be applied to Norco's small population, and former CCN leader Margie Richard asked whether Shell used "Method 21" to find leaks at the plants. Questions explicitly about health were no more numerous: Ramachandran posed a question about the short-term effects of chemical exposures, and another, white man wanted to know about the relative contributions of indoor and outdoor air quality to chemical exposures. But health was also just below the surface of Audrey Taylor's inquiry into how often the Louisiana ambient air standards were updated: she wanted to know whether they might be out of date with respect to current knowledge about how chemicals affect health.

The personal experiences, opinions, and concerns that motivated these questions were conspicuous in their absence. Audrey Taylor did not say, "I believe you that you are in compliance, but still my family is sick with things that the doctor told me have to do with living around these chemicals, so could there be something wrong with the standards?"—a formulation that had come through clearly in an interview I had done with her the prior week. Margie Richard, likewise, did not mention her asthmatic grandson, hospitalized twice before the age of ten, as a reason for her concern about Shell's methods for controlling its emissions, as she had throughout CCN's campaign for relocation. The personal or experiential claims that did creep into the questioning appeared in a way that distanced them from the speaker. One black panelist commented that New Sarpy had been completely forgotten in the study, citing a recent incident at Orion as a reason why New Sarpy needed monitoring even more than Norco. What he did not acknowledge was that he himself lived in New Sarpy; instead, he framed his remark in such a way that it could have come from anyone on the panel. The parish council member prefaced his question about short-term health effects by making reference to a local knowledge claim popular among white Norco residents: air pollution cannot be making people sick, the claim went, because everyone who lives around *me* lives into their eighties and nineties. In the context of the Community-Industry Panel discussion, however, the assertion was stripped of its experiential aspect and re-presented as collective, rather than individual, truth. "We are confident," he said, that there's no long-term effect from this because "we know" that people in "our community" live to be eighty and ninety years old.[28]

Framed in this generic way, residents' comments gave industry members and other experts in attendance the opportunity to elaborate on technical issues and processes in a way that advanced their favored ways of framing issues. For example, the study's conclusion that Norco's air was uniform was, in fact, a confirmation of experts' hypothesis: if chemicals were being released from high stacks and air in the small town was relatively well mixed, then, according to their computer models, Norco's air *should* be uniform. As a result, statistically insignificant variations across the six monitoring sites were taken to be indications of ground-level leaks or other, so-called fugitive sources. In this context, questions about pollution sources—natural versus man-made; highway versus industry—became occasions for Shell and Motiva engineers to talk about their methods for tracking down what they considered anomalous chemical levels. In particular, in response to a question about higher levels of chemicals at a monitoring station near a busy highway, a pair of Shell's engineers explained how they were using "grid sampling" to pinpoint the source of the excess emissions, suspected to be from a set of storage tanks across the highway from Shell's facility, and how they had already located a leaky valve in an ethylene pipeline using that method. Interestingly, in these particular interactions, industry engineers resisted community questioners' invitations to shift the blame for chemical levels away from industrial facilities to, for example, traffic. However, in telling how they sniffed out the causes of unexpectedly high chemical concentrations, they also reinforced their problem-solving framework and refused to acknowledge the possibility that ground-level leaks and other unplanned releases are a systemic problem—a way of understanding industry's environmental effects developed by organizations like LABB and employed in community campaigns.[29]

Audrey Taylor's question about when the Louisiana ambient air standards had last been updated similarly became an opportunity for experts to reassert the appropriateness of comparing air quality data to regulatory standards. Luann White assured her that standard setting was an ongoing process, in which the state revised the standards every time they got new data about how health effects were related to chemical concentrations.[30] By upholding the authority of the LDEQ's standards, White defended the decision made by the technical team of "Air Monitoring…Norco" to use the standards as a framework for assessing air quality—and deflected both Taylor's unspoken skepticism and environmental justice activists' charges that standards are *in*adequate to protect health.

Notably, it was the panel's prohibition on personal experiences—a prohibition stemming from dialogue's particular implementation of deliberative ideals—that allowed White and other experts to advance their preferred frameworks, and that insulated them from the most pointed critiques. Environmental justice activists' most successful criticisms of scientific understandings have been built upon residents' experiences; by pointing to community members suffering from disease at rates that current science does not explain, they have justified their calls for alternative ways of knowing. The success of these challenges has depended heavily on community members' testimony, on residents of fenceline communities saying, "If the refinery is so benign, then why is every member of my family sick?"—on, that is, precisely the kind of personal, interested comment that CAPs and the Community-Industry Panel, through their particular structures of deliberation, consider outside the bounds of reasoned dialogue. Without the possibility of personal testimony, experts cannot be called to account for the limitations of their frameworks; Luann White does not have to explain why, if all regulatory standards used to evaluate the performance of Shell and Motiva and Orion are being met, Audrey Taylor's brother has a tumor and her daughter died of cancer.

The kind of local, experience-based knowledge on which activists' challenges to expertise are based can be transformed into general claims acceptable in the context of community-industry dialogue. Ramachandran, in fact, did precisely that in advancing white Norco residents' individual observations that their neighbors were living to a hardy old age as something "we know" about health in Norco. His claim went unchallenged at that particular panel meeting, no doubt partially because experts, who did not actually acknowledge the assertion, had no interest in disputing it.[31] But generalized, local knowledge–based claims challenging to those of industry engineers and scientists did occasionally surface in CAP and Community-Industry Panel meetings. In these cases, the deliberative structures of the panels enabled industry members to undermine the claims and deflect critique without appearing heavy-handed.

In 2001, Greenpeace activists allied with CCN made their presence felt in St. Charles Parish. An industry member responded to their claims that the industrial corridor was a "cancer alley," during a meeting of the St. Charles CAP:

> [Stanley Dufrene] also advised that Greenpeace is back in Louisiana. Greenpeace had a photo display with very inflammatory messages— "Cancer starts here," "Cancer Alley," etc. Mr. Dufrene reminded all present that based on Dr. Chen's presentation to the CAP last year, Louisiana cancer rates are actually at or below the national level. The emotional sound bite headline used by activist groups and the media has no basis in fact.

Presented as an announcement and not generating any further discussion, Dufrene's comment is presented as a helpful reminder of information presented by an outside expert. In formulating it this way, he draws on his status as an equal member of the CAP who had, like resident members, been in the audience for Chen's presentation—rather than on any special knowledge that he himself might have. Dufrene thus dispensed with activists' knowledge claim in the same manner in which industry members steered other aspects of the CAP: by offering a helpful suggestion from his structurally equal position. He also capitalized on industry members' ability to shape the discussion to reflect experts' frameworks. He was able to correct emergent misperceptions without violating the egalitarian premise of the meeting in large part because he could refer to the Louisiana Tumor Registry's approach to understanding chemical exposures and health effects, an approach embraced by industrial facilities.

The CAP's way of instituting the deliberative restriction on particular interests also helped Dufrene's comment go without challenge. Aside from

Greenpeace's "Cancer Alley" assertion—arguably a knowledge claim based on the aggregate local experiences of members of multiple communities near industrial corridor chemical facilities—resident members who might have been skeptical of Chen's conclusions had no comparable, generalized knowledge to advance in response to Dufrene. No scientific studies adequately accounted for some CAP members' observations that cancer was suspiciously present in their families or neighborhoods, and the CAP's norm of excluding members' specific circumstances from discussion prohibited them from responding with their own experiences of cancer in their communities.

Community Advisory and Community-Industry Panels, then, are clearly shaped by industrial facility representatives' interest in conveying favorable messages about their companies and scientific understandings that absolve industry of responsibility for environmental degradation or ill health in neighboring communities. They are not, however, simple exercises in advertising or brainwashing; rather, they do embrace the deliberative ideals that industry's rhetoric of "dialogue" suggests. Indeed, it is these very ideals— or, rather, the panels' implementation of them—that allow industry representatives to establish the authority of their technical understandings and, simultaneously, to satisfy community members' desire to be treated with the respect due equals. By presuming the equality of all panel members, these community-industry dialogues mask industry members' power to shape the content of deliberations. Similarly, by presuming that individuals' particular circumstances and stakes are not relevant to deliberations on issues of general interest, they structurally exclude the local knowledge that underpins the most trenchant and effective critiques of industry expertise.

The deliberative structures of the CAPs and Community-Industry Panels, further, were consequential for community-industry relations more generally: they allowed industry engineers and scientists to retain—or, in the wake of campaigns such as that in New Sarpy, to reestablish—their authority over scientific issues without overtly putting themselves above community members. In negotiations in Norco, Shell engineers could only keep their claim that their operations did not cause health problems out of the discussion by declaring the authority of their science and reminding CCN members of their power not to negotiate at all. In contrast, in community-industry dialogues, industry representatives can forestall critical examination of their claims about the health effects of chemicals without disrupting the egalitarian tone of the panels—while, that is, continuing to treat community members like equals. And in the process, experts go from being know-it-alls who will not listen to community members to being helpful, down-to-earth guys whom one can not only talk to but relate to as well.

Conclusion: Deliberating Expertise

At the American Association for the Advancement of Science (AAAS) in the summer of 2009, I gave a lecture titled "What Citizens Know That Scientists Don't." Using bucket monitoring by New Sarpy residents as one example of the way nonscientists formulate legitimate alternatives to scientific ways of knowing, my talk was an attempt to get community groups' challenges to expertise taken seriously in environmental policy making. Asked by one of the science policy professionals in the audience to explain just how citizens and their local knowledge could be included in decision making, I fell back on the idea of deliberation as a vehicle for reevaluating scientific ways of knowing. I suggested that, for example, standards for environmental monitoring procedures should be set through discussions that included citizens as well as scientists.[32]

But was my faith in deliberation—the same faith expressed in efforts by numerous STS scholars to analyze, influence, create, and reimagine public discussions on controversial scientific issues;[33] the same faith, in fact, represented in calls by CCNS members and their activists allies for talks to resolve tensions in New Sarpy—was that faith justified? Despite key deliberative aspects, residents were able to question expert knowledge neither in community-industry dialogues like CAPs nor in activist-initiated negotiations like those in Norco. And even public forums specifically designed to be deliberative have been shown to subtly reinforce the authority of science rather than opening it up to fundamental critique.[34]

A better answer to the question at the AAAS—an answer that took into account what I knew of community-industry discussion in St. Charles Parish—would not have invoked deliberation so glibly, as an answer in itself. Instead, it would have specified the kind of deliberative process necessary for incorporating citizens' knowledge and ways of knowing into policy. That is, the contrast between negotiations in Norco and community-industry dialogue on CAPs demonstrates that the effectiveness of deliberations—in overcoming power differentials; in fostering reasoned discussion of the common good; in making science and scientific ways of knowing a subject for critical debate—depends not on the embrace of deliberative ideals but on the specific ways in which those ideals are implemented. How is equality among participants achieved? What conditions must their comments satisfy in order to qualify as something more than a statement of particular interests? Rather than merely calling for citizens to be involved in discussions of scientific standard setting, then, I might have said that citizens should be included in deliberative processes that considered their local knowledge claims, and the personal experiences that grounded them, to be legitimate contributions to the general discussion; that provided for the participation of environmental justice organizers, STS scholars, and other professionals with the training and experience to help citizens frame their particular experiences as more general claims to knowledge; and that were accompanied by mechanisms, potentially including social movement organizing, that compelled decision makers to act on the results of the deliberation.

Although not usually made explicit with respect to deliberations on technical issues, the structures for discussion that I might have suggested—that I would, in fact, advocate for—are in keeping with the recommendations of liberal theorists concerned with representing the voices of marginalized people in deliberative democracy. They, however, run counter to *neo*liberal trends in environmental governance. Regulatory agencies' interest in seeing environmental justice disputes settled through community-industry dialogue— an interest shared by the petrochemical industry—not only intensifies the problem of equality among participants. By eliminating state participation, it also distances deliberations from real influence on environmental decisions and actions. That is, citizens can influence their government through a variety of mechanisms, and many regulatory agencies sponsor public discussions explicitly to inform their programs and policies. But, outside of community organizing, which community-industry dialogue is clearly meant to replace, it is far from clear what mechanisms will allow citizens to influence the behavior of corporations. Reflected in the CAPs' presumption of equality among clearly unequal participants, the tendency of neoliberal policies to place individuals, rather than social structures, at the center of politics also works against deliberative practices that would foster critical discussion of technical issues. By refusing to acknowledge that residents are at a disadvantage with respect to industry experts, community-industry dialogues rule out the possibility of including other professionals who could help decrease the gap between experts and community members' ability to engage scientific ways of knowing and translate technical claims into reasoned argument about the general good.

And yet, despite the ways they diverge from a structure that could enable nonscientists to challenge expert knowledge and ways of knowing, community-industry dialogues informed and promoted by neoliberal policies have been compelling to residents of New Sarpy and other fenceline communities precisely because of the deliberative norms they implement. The panels' presumption of equality and focus on the general good did more than let industry experts exert their scientific authority without making their greater knowledge and power explicit. The panels' deliberative structures also let

CCNS leaders, criticized by fellow citizens for their contentious, disruptive campaign, appear once again as reasonable, civilized people and regain their claim to "building up the community" by representing their neighbors in discussions of the common good. And if community-industry dialogue circumscribed residents' ability to challenge the industry's technical claims—an aim that even the successful campaign in Norco did not ultimately achieve—it did so in a way that allowed community members, even those convinced of industry's health effects, to feel as though they were being "treated as equals."

One among a suite of neoliberal practices for environmental governance, community-industry dialogue functions to shore up expert authority against attempts to democratize scientific knowledge. It satisfies demands for egalitarianism, for civility, for reason around contentious issues of pollution and health in fenceline communities. Yet it does so by bracketing out the situated, experiential knowledge that represents the most serious challenge to experts' ways of knowing, and that democratizers are particularly keen on seeing become a systematic part of environmental decision making. In the neoliberal guise of dialogue, deliberation thus becomes a resource for the reassertion and maintenance of expert authority over questions of the environmental and health impacts of industrial emissions.

5

Responsible Refiners

Jack Stanley has been trying to build a refinery [in New Sarpy] since the midseventies. He has had a reputation of shortcuts and . . . it's a long, convoluted history of layoffs, poor management of people, not good engineering, just poor practices. There's a lot of horror stories, whether it's fact or fiction, in the industry about some things that have happened in the past.
—Jason Carter, Orion Refining, May 22, 2003

I lived on Apple Street when I was a kid, and I remember Good Hope [refinery]. . . . I was not too happy when I heard I was being sent here. But I thought that Valero could do it. That is, if anyone could do it, Valero could.
—Ellen Williams, Valero St. Charles Refinery, February 10, 2006

As it happened, Orion and I moved out on the same day. On June 30, 2003, as I set out to drive back to California, the refinery that had been Orion became the Valero St. Charles Refinery, one of over a dozen facilities owned by North America's largest refiner, Valero Energy Corporation. The change of ownership was a good thing for the community, Orion managers had assured residents at the previous month's Community Advisory Panel (CAP) meeting. Valero was a big company with a lot of resources; it had allocated $400 million over the next five years for improvements to the facility. It had readily agreed to take over the home improvement loans, cash payments, beneficial environmental projects, and other community improvements that Orion had promised to residents and regulators, and, beyond that, it had a history of being very involved in the communities where its refineries were located.

On my next visit to New Sarpy in February 2006, it appeared that Orion officials had been right. As I made the rounds of the St. Charles Terrace neighborhood, visiting with residents who had become friends over the course of Concerned Citizens of New Sarpy's (CCNS's) campaign, they told me over and over again how much of an improvement Valero was over Orion. Under Valero's management, the refinery looked better than it ever had before: where the drop yard had been, according to Harlon Rushing, you wouldn't even find a piece of paper lying on the ground. "They listen to us," Myrtle Berteau said, describing how easy it was to get Valero officials to come to community meetings—not that they usually needed to, since Valero invited them to meetings on site as well. She and her daughter-in-law also raved about how nice they were, in contrast to Orion's management. Even Ida Mitchell, who was disgruntled that the refinery's beneficial environmental projects had not included culverts for the neighborhood, declared that the folks at Valero were "a different cut of people" than those at Orion.

Even more telling was what residents did not say. In my ten-day visit, I heard not a single complaint about accidents, or flaring, or smells from the refinery—topics revisited daily during Orion's tenure. And, while I learned from Ron Guillory, Valero's middle-aged black director of public affairs, that the refinery was in the process of building two new operating units to meet low-sulfur fuel requirements, residents scarcely mentioned the new construction, nor did they oppose the company's application for a permit for a $900 million expansion that would take the refinery's capacity from 260,000 to 350,000 barrels of crude per day. Even in the Valero CAP meeting that I attended during my visit, the agenda-setting discussion included only brief references to "environment" and "odors"; instead, community outreach, emergency preparedness, and the strain that two thousand new residents had placed on the parish's water system took center stage.[1] Residents' sense that Valero's staff were a cut above apparently extended to the refinery engineers' ability to prevent the kind of environmental impacts that CCNS members had repeatedly criticized Orion for.

Valero officials, of course, knew that they had immeasurably improved relations between the refinery and its neighbors; indeed, they'd been trying to do exactly that. According to Ron Guillory, Orion had been so bad in terms of both the way it ran the refinery and the way it treated its neighbors that Valero could hardly help but do better. But health, safety, and environment manager Ellen Williams, a white woman in her forties, refused to blame Orion's problems on the level of niceness, integrity, or even technical competence among its staff. She suggested that she and her colleagues were able to be good neighbors not because they were necessarily of a different cut but because they were part of a different kind of company. In Williams's view, Orion, and Transamerica before it, had gotten into trouble because the companies were "not sophisticated enough" to know how to handle their environmental issues from a technical or social standpoint:

They weren't true, long-term refiners in a big company that understood how those, what seemed to be small things, could have such an impact on a community. They had a tough start-up, they flared a lot, they didn't have the resources to help them talk with the community, to understand that importance, you know, they were just about getting the refinery running so we can sell it and make a lot of money. It's not that they didn't care about the community. . . . I don't think that's, that there was some moral issue there with Orion, I never got that feeling. . . . They didn't have this corporate culture that we, it's just big value to be a positive influence in a community.

Because refining was at the heart of Valero's business, Williams went on to explain, the company was interested in having its St. Charles refinery become a model of "safety compliance, environmental compliance, and profitability. You know, those important things to Valero." Those corporate values were backed up by technical competence, of course—in order to be a "positive influence" in the community, in Williams's view, a refinery had to operate reliably, and operating reliably required good designs and good maintenance. But it was Valero's values, more than the technical abilities or moral stature of any of Orion's managers, to which Williams attributed New Sarpy residents' new confidence in the refinery's operations since Valero assumed control.

In both community members' and Ellen Williams's accounts of the refinery's transformation, then, scientific authority and the morals or values of technical practitioners were tied together: where Ida Mitchell, having nothing to complain about in terms of Valero's environmental performance, referred to the moral fiber of Valero's manager in calling them a "different cut of people," Williams talked about corporate values as central to the way she and her staff did their work. That technical and moral status would be associated is little surprise. Mid-twentieth-century sociologist of science Robert Merton posited the connection in his argument that the authority of scientific knowledge stems from virtues, among them skepticism and disinterest, internalized by members of the scientific community.[2] More recent scholarship has demonstrated the connection through empirical case studies from a variety of historical periods, showing how scientific credibility hinges on the scientist's moral standing as, for example, a "gentleman" in the seventeenth century or an "authentic" person at the turn of the twenty-first.[3]

What is unusual in this case is that Williams's account associates technical credibility and competence not with the moral stature of individuals but with the company for which they work and *its* moral standing. Typically, being associated with a large corporation is considered a detriment to scientific credibility: in the context of numerous, high-profile cases of companies and industry groups suppressing data that shows their products in an unfavorable light and manufacturing uncertainty about mainstream scientific conclusions with potential negative consequences on their business, scientists employed by industry are hard pressed to refute accusations of bias.[4] And for engineers, while corporate affiliation tends to be taken for granted—in the United States, modern industry and the engineering profession co-evolved in important ways[5]—it is professional societies, not corporate employers, that are the source of engineers' ethical codes. In fact, at the heart of many case studies in engineering ethics is a tension between the engineer's obligation to her employer and her obligation to the public good.[6]

The idea that technical integrity and corporate employment are at odds rests on the assumption that corporations are driven only by profit. But large companies are increasingly challenging that assumption by publicly establishing corporate codes of ethics that serve as governing principles for the organization—and, in the process, fashioning themselves as moral actors, voluntarily committed not only to fiscal responsibility but to social responsibility as well.[7] Itself an outgrowth of neoliberal trends toward deregulation, globalization, and, in sociologist Ronen Shamir's term, "responsibilization,"[8] the idea of corporate social responsibility (CSR) enables companies like Valero to argue against prescriptive government regulation of their activities:[9] as ethical actors, they are capable of regulating themselves and doing the right thing without coercion—or so the logic goes.

Indeed, in New Sarpy, the takeover of the refinery by a responsible multinational company did improve its community relations and environmental performance without any change in the regulatory context.[10] According to Ellen Williams, it was her company's values that allowed her and her staff to make the plant more reliable and to ensure that the company would be a good influence in the community where Orion personnel—whom Williams credited with good intentions—had been unable to. But the case of New Sarpy shows that the idea of the corporation as a moral actor has deeper consequences, as well. In their interactions with community members, engineers and scientists like Williams were able to refer to their companies' ethical commitments to help bolster their own personal claims to being responsible, competent technical professionals—and thus better maintain their moral status even when accidents and releases occurred at their facilities. Further, in the context of a technically oriented industry, corporate *social* responsibility created a realm in which plant managers had obligations to perform well, but in which they claimed no expertise. When community conflicts arose, this social realm was divided from the technical—even though fiscal, environmental, and community responsibilities were in many cases regarded as seamless—and community grievances framed as social issues, requiring thoughtful attention from managers, but no rethinking of technical practices. By taking on social responsibilities as part of their core business values, then, petrochemical companies created a space for plant managers to admit serious faults in their interactions with residents, and thereby resolve community conflicts, without jeopardizing their technical authority. Neoliberal commitments to CSR thus became one more resource that industry engineers and scientists drew on to establish, and reestablish, themselves as credible, competent experts on contestable issues of environmental quality and health.

Sites of Social Responsibility

When I next encountered Ellen Williams in 2011, her face smiled out at me from the top right-hand corner of Valero's home page.[11] In the five years since I had interviewed her, the caption informed me, she had risen to the position of vice president of occupational and process safety, and she was featured as one of a rotating cast of Valero presidents and vice presidents under the headline "Excellence in All We Do." Their short statements—Williams's was "Safety is our No. 1 Priority"—served as a gateway to a page entitled simply "Excellence," where each discussed various aspects of the company's "Commitment to Excellence."[12] Spelled out in detail on its own page, the Commitment to Excellence encompassed five separate commitments, only one of which referred to Valero's money-making mission. Even at that, the "Commitment to Our Stakeholders" promised not to maximize profits but to "deliver long-term value" to investors, employees, customers, *and* communities through "profitable, value-enhancing strategies with a focus on world-class operations." The remainder of the Commitment to Excellence embraces other, apparently less self-interested, goods, including safety—which is literally first among the commitments, outranking the commitment to stakeholders—employee satisfaction and fulfillment, environmental quality, and community involvement.

Referred to obliquely by Williams in her 2006 description of her company's values, this Commitment to Excellence marked Valero as, in her words, "a responsible company." Corporate codes like the Commitment to Excellence are central to corporate social responsibility, which, though varied in its implementation, is at its core a model for doing business in which a company acknowledges that its obligations extend beyond its investors.

Large corporations embraced CSR, in its contemporary form, in the 1990s as a response to a "crisis of legitimacy" that had been building for two decades: as neoliberal policy makers removed barriers to international trade, they also rolled back many of the regulations thought to keep profit-hungry corporations in check. The combination of globalization and deregulation led to public concern about the possibility of rampant human rights abuses and environmental despoliation by companies of unprecedented size and power.[13] In the petrochemical industry, these concerns were compounded by high-profile accidents at chemical plants, most notably the deadly release of over forty tons of methyl isocyanate from a Union Carbide plant in Bhopal, India, in 1984.[14]

Now standard among multisited, multinational petrochemical companies—Shell and Motiva, for example, share a Statement of General Business Principles that grounds their principles, including safety, environment, and community well-being, as well as long-term profitability, in "core values" of honesty, integrity, and respect[15]—corporate codes of ethics and statements of social commitment are themselves a distinctly neoliberal response to this crisis of legitimacy.[16] They allow corporations to volunteer to be accountable to social norms rather than being held accountable through prescriptive government regulation, and they are "enforced" by transnational social movements and nongovernmental organizations—the same groups that pressed companies to adopt them in the first place—who work to rally public opinion against companies that violate accepted or professed standards of ethical conduct. As a result, CSR tends to be embraced by very large corporations, usually with national or international brands to protect;[17] the kind of border-spanning activism capable of influencing the practices of a company like Shell, for example,[18] would be neither possible nor effective where a company like Orion was concerned.

Voluntary commitments to social responsibility are able to stand in for government regulation of multinational corporations because they position companies as moral actors, capable of governing themselves.[19] Implicit in corporate value statements is the claim that the integrity at the core of Shell's business activities, or Valero's ongoing quest for excellence in all it does, would in itself lead the company to do the right thing without having to be told to do so by some outside authority. In these statements, however, corporations' capacity for and commitment to principled action is never in tension with its profit motives. Rather, CSR suggests that doing the right thing is actually good for business—solidifying corporations' claims to moral agency by integrating ethical obligation into the very logic of the market.[20]

As a feature of extensive, multisited corporations, CSR demands and relies on the responsible action of workers at all levels of a company.[21] This is explicit in Shell's Statement of General Business Principles, which stresses the need for employees (in contrast to management) both to act in accordance with the principles and to report breaches of them, but it is also reflected in Valero's *Social Responsibility Report*, which shows "how the company puts into action its Commitment to Excellence" in large part through reports on site-level activities. In the Environment section of the report, for example, the St. Charles refinery is touted for "its environmental justice leadership" and for having gone six years without an acid gas flaring incident—the kind of incident that had been so prevalent at the refinery during Orion's tenure. A number of sidebars even feature the contributions to excellence made by committed individuals, including a reliability manager who started a preventive maintenance program for operators at one refinery and a rank-and-file worker at another refinery who identified and fixed a series of faulty electrical components whose failure would have jeopardized operations.

In St. Charles Parish sites of "responsible companies," high-ranking managers like Ellen Williams clearly incorporated their companies' stated commitments into the way they conceived their professional responsibilities, into their goals for their plants, and into their decisions about what activities and programs to pursue. Although managers did not usually invoke their corporations' codes explicitly, notions of corporate social responsibility provided a *business* rationale for community programs and investments in environmental performance. At the same time, the intertwining of community, environment, and profit enabled by the logic of CSR offered managers a way to set limits on their community and environmental activities and to justify their refusal of certain demands by activists.

Value for Community

Just as corporate statements of commitment, and CSR more generally, incorporate rather than reject profit-making aims, St. Charles Parish plant managers talked about their responsibilities and goals in ways that integrated obligations to the community, the environment, and the bottom line. Each plant manager I interviewed told me that his[22] primary responsibilities were to deliver products and to make sure that his facility was profitable—but those whose plants were sites of large companies quickly connected those aims to other obligations embraced by their corporations. Shell Norco's plant manager, Wayne Pearce, for example, wrapped profitability, environmental performance, and social responsibility together by saying,

I operate this facility in a safe and environmentally sound manner, in order to deliver the things which deliver economic benefit to Shell and to the community around us. The primary driver in terms of economic benefit is economic benefit to Shell. We're a capitalist company in a capitalist Western world, so we make profits. But there are spinoff benefits in any economy where you've got a company operating. So that's what I deliver.

In Pearce's description of his charge as plant manager, environment, profits, and community were inextricable. In keeping with Shell's principle of treating health, safety, security, and the environment as "critical business activities," being profitable depended on being safe and environmentally responsible. And by mentioning "spinoff benefits," which subsequently led to a description of the community programs that he saw as part of his plant's mission, Pearce suggested that profitability was at the core of any social good that his plant could do—echoing Principle 1 of Shell's General Business Principles, which reads in part, "Without profits and a strong financial foundation, it would not be possible to fulfill our responsibilities."

Understanding social and environmental responsibility as central to the business of running a refinery—as Shell, Motiva, and Valero all did— allowed plant managers to see as part of their mission addressing the needs and concerns of a variety of stakeholders who were neither consumers nor investors. Among the most prominent of these stakeholders were plant employees: in their interviews with me, managers mentioned their obligation to create a workplace that valued diversity, allowed employees to balance work and family commitments, and provided opportunities for skills development and advancement. Neighboring communities, too, were cited as especially important stakeholders; in fact, Shell Norco's Good Neighbor Initiative manager, David Brignac, explained to me that neighbors were the real source of a plant's "license to operate," an idea echoed by plant managers.[23] In describing how they thought about their obligations to community stakeholders, plant managers echoed the general commitment to the well-being of the community described by Ellen Williams when she said that Valero had a "value for being a positive influence in the community." According to Mitchell Mobley, it was important that his plant be "donating time and money to making this place a better place to live," so that neighbors could see that the company "cares for the community"; Wayne Pearce, similarly, said he had to be "thinking about how we impact local society for the good."

What it meant in practice to have a positive impact on a neighboring community, managers were quick to point out, required judgment on the part of plant managers and could vary significantly from place to place. As Mobley, who had managed several refineries, described it,

> There isn't a recipe, Gwen, there is not a recipe. What feels good in [Richmond, California], in [Wilmington, Delaware], in [Baton Rouge],[24] in Norco, may all be different. And it's as you work together with the community, I think you get to the point where the company feels they're doing what they can, the community generally feels they're doing what they can and you live happily ever after. Some places it takes more than other places, obviously.

Community expectations of neighboring facilities varied not only across the United States, as Mobley suggested, but from country to country as well. According to Wayne Pearce, who began his Shell career in the United Kingdom, in the United States, "the expected role of industry in terms of support to education or local government is at a higher level than it is in a country that's got a more social[ist] government." The variation in public expectations of petrochemical plants left plant managers with the task of figuring out what kind and degree of community engagement was most appropriate for their facilities.

But although the corporate codes of multisited, international companies could not specify how their values for "being a positive influence" in a community were to be implemented in a particular place, the codes' larger framework was a resource for plant managers trying to decide what community programs they should direct resources to. In "making choices between supporting the local ballet and supporting the local education system," for example, Pearce told me that what he considered was, "what's most appropriate, what fits best in terms of our goal to be here for the long term." The goal invoked—without parroting—Shell's General Business Principles, in which long-term profitability, continued growth, and sustainability, conceived as having economic, social, and environmental dimensions, are prominent. And with that goal in mind, Pearce chose to support community programs that not only met community needs but made sense in a framework that saw profitability and commitment to community as mutually reinforcing. He gave the example of the recently launched Community Education Initiative, a Norco community program heavily underwritten by Shell that not only offered GED classes for adults wishing to get high school diplomas but combined them with educational programs for young children, as well as transportation for participants, in order to make the GED classes more widely accessible. Pearce was clearly proud of the program and its unique combination of adult and early childhood education, an innovation suggested by community members, yet he also stressed that it was central to his company's core aims: "It's not altruism or anything like that, having well-educated people is the feed to the workforce that we need for the future. So you can wrap all those things together in this, we're here for the long term." The educational program, then, made sense in the context of Shell's core principles; namely, it was consistent with a framework for understanding corporate social responsibility that made commitments to community and workers integral to the commitment to profits. Indeed, CSR's logic of social responsibilities as intertwined with profits helps explain the prominence of educational programs among petrochemical plants' community activities; contributions of a similar scale to the local ballet would have been unusual and much harder to account for.

The logic of CSR not only guided plant managers in deciding what sort of community programs to engage in; it also helped managers justify decisions not to engage. For example, after Shell in 2000 announced its Voluntary Purchase Program—which offered a premium for properties on the two streets closest to the fencelines of Shell's east and west sites—they resisted calls from Concerned Citizens of Norco (CCN) to extend the program to Diamond's other two streets on the grounds that extending the offer would be bad for the community, and thus not in keeping with Shell's commitments to Norco. Speaking with Steve Lerner during that period, David Brignac explained that the company had wanted to create a buffer between its Norco plant and the community, of a size that encompassed the two streets. But, Brignac said,

> If we establish that these are the areas we want as a buffer, and we extend the buy-out in Diamond, we don't have good logic to do that. We don't need it for a buffer so why would we buy it? For some other reason. We can't get an understanding of how we can rationalize it and we think that if we do [buy] it very quickly we will say, well, we expanded it on this side, in fairness of the community we need to expand it on this side. . . . So now you are buying out such large chunks of Norco that you are really threatening the integrity of the town. By buying out hundreds and hundreds and hundreds of residents you start getting a lot of opposition . . . not just from people in Norco saying you are buying out too much property but you have schools here, you have businesses, you have politicians . . . you are creating a problem that you don't have to create, basically. So we think it is not good for Norco, from that standpoint.

According to Brignac, Shell's resistance to buying out the rest of Diamond was, again, based on interlocking principles core to the company's values. The implication is that, while having a buffer zone made sense—as a way of minimizing the noises and odors perceptible to neighbors, and thus helping to secure the plant's license to operate—having such a large one would not make the plant less obtrusive or more acceptable, making it impossible to justify the expense of buying out more homeowners. More overtly, Brignac argues that extending the program would actually be at odds with Shell's commitment to the community or, in his words, their "genuine intention to want to do what is right for Norco and for Diamond."

High-level corporate values, interpreted locally, thus gave plant managers a rationale for rejecting some community expectations and demands, even while responding favorably to others. At the same time, the multilevel nature of corporate commitments offered managers an opportunity to refuse certain kinds of engagement without having to categorically dismiss them. In particular, plant managers were loath to deal with the environmental activist groups that lent their support to community campaigns. In addition to questioning activists' intentions (one high-ranking manager explained to me that activists were necessarily interested in prolonging, rather than settling, fights in order to justify their own existence) and their methods (another told me that an activist group's decision to politicize a colleague's choice of residence sickened him), they accused environmental activists of obstructing their attempts to be a positive influence in the community. Orion officials, for example, blamed the Louisiana Bucket Brigade (LABB) quite overtly for the protracted period of contention with CCNS: LABB had riled residents up, incited hostility at public meetings, and prevented Orion from speaking directly to their neighbors.[25] But from a site of a company with an expressed commitment not only to the neighboring community but to society in general, Wayne Pearce framed the issue rather differently. Asked whether environmental groups were among his stakeholders, he replied that "the face of environmental activists" was, for him, Concerned Citizens of Norco—the group of Diamond residents campaigning for relocation. Through them, he came into contact with groups like LABB and Greenpeace; however, "The legitimate discussion about environmental matters that takes place in terms of engagement at Norco is via CCN and other interested groups. The legitimate engagement of Greenpeace at a national level isn't me." Pearce thus articulated a neat division of labor: where "communication and engagement"—Shell's Principle 7—were concerned, his responsibility was to locally based groups; someone in Shell's corporate offices executed that responsibility with respect to extra-local groups like Greenpeace. By dividing responsibility in this manner, plant managers at multisited corporations were still able to refuse the intrusion of outside groups—but without being overtly hostile to environmentalists or environmental causes.

Investing in the Environment

Corporate commitments to social responsibility thus let plant managers not only make contributing to the community a priority but also set limits on the nature and extent of their responsibilities. They operated similarly with respect to facilities' goals of "operating in a safe and environmentally sound manner." In the strictest sense, of course, community obligations and environmental obligations are structured quite differently: any sense of obligation that a petrochemical facility might have to a neighboring community is constructed entirely as part of a voluntary commitment to corporate social responsibility; in contrast, state and federal laws mandate environmental and safety benchmarks that petrochemical facilities must meet, and limits on each facility's emissions are set by an operating permit negotiated with and granted by state environmental regulators. Yet, when high-ranking managers at St. Charles Parish plants talked about their environmental goals and responsibilities, they did so in a way that subsumed regulatory compliance within the larger framework of CSR. For example, John King, the white, thirty-something environmental manager at Motiva Norco Refining, told me that it was his job to make sure that the plant complied with environmental rules—but that he and his group were also concerned with making improvements beyond what the regulations required:

> We're working to improve, and not just improve in terms of being more perfectly legal, but improve in terms of going beyond the minimum required by law. And that's partly my responsibility, but that, I think, is spread more across the plant in terms of everyone has that personal responsibility because that's a corporate value that [Motiva] has.

The commitment to environment, in King's characterization, included compliance but was not limited to it; rather, because environmental performance was a "corporate value," King and his team established "environmental management systems" that asked every employee to assess the environmental impacts of their day-to-day work and look for ways to ameliorate them. Similarly, Shell Norco's Health, Safety, and Environment manager Randy Armstrong described himself as responsible for ensuring that people don't get hurt, that we don't have serious incidents, fires and explosions, and that we comply with the laws of the land, and that we make progress on reducing our environmental, or as it's becoming our social, footprint. In the communities that we're part of.

Armstrong's description of his duties, like King's, paired compliance with a notion of self-directed improvement. Moreover, he tied environmental and social responsibilities together (albeit elliptically), suggesting that minimizing environmental impacts was somehow integral to making good on commitments to nearby communities.

Seeing environmental improvement—not just environmental compliance—as part of the larger fabric of social responsibility, petrochemical facility managers conceived their environmental goals largely in response to expectations of the various stakeholders acknowledged in their CSR statements, especially neighboring communities. Community members, according to plant manager Mitchell Mobley, expected a refinery not to be a nuisance, which he interpreted as, "no smoke, no smell, no noise . . . at simplest terms, we want to be invisible to the community." This expectation led him to address site-level issues that were not covered by environmental regulations: he told me how he had been appalled to be told by a community member that something from the plant would occasionally cause her home to vibrate in the middle of the night. Once he was able to track the vibrations to a compressor system that had "had the shakes" since it was installed, Mobley told me, "I told our environmental manager, 'John,' I said, 'we've got to do something about this. I don't want neighbors to tell me this thing is shaking and it's normal to them.'" Similarly, Randy Armstrong explained how community expectations led to his plant's decision to cut flaring by 50 percent over a three-year period:

> We got invited to a group by [EPA's] Region 6 [office] to take a look at episodic releases. And part of it was their opinion, driven by the community concerns that they got, since their second highest complaint area was flaring in Region 6, at least according to Sam Coleman, it was, when he brought us all in, was flaring is unacceptable. [paraphrasing Coleman:] I don't care what the modeling data that you have and the fact that you view this as a safety device and that when the flares go off, you actually feel like, well, you got all your safety devices working. It's unsightly, and it's no longer acceptable to have the level of flaring that you do.

Although no regulations changed to outlaw or penalize flaring—which, as an episodic release, is not counted toward permitted emissions limits—communities in Louisiana and Texas, among others,[26] had complained enough about the practice to establish it as undesirable environmentally. Conveyed to industry by regulators, the shift in public opinion pushed Shell—and possibly other companies—to decide that eliminating the failures that lead to flaring was the responsible thing to do.

Prioritizing environmental performance costs money, and environmental managers indicated that part of their job was to persuade their organizations to invest in what they deemed to be necessary environmental improvements. For example, having heard from the EPA that the agency no longer found the industry's practices of flaring acceptable, Randy Armstrong said that he had to go back to Shell and get commitments from his management to reduce flaring. In this kind of task, corporate commitments were clearly an important resource for environmental managers: Armstrong suggested that he was able to secure a commitment to a 50 percent reduction in flaring because he was able to convince Shell that the issue was one that people—community, regulators, environmental groups—cared about. Similarly, Ellen Williams said that she had been able to shut down and fix the two sulfur recovery units that had caused all the flaring in New Sarpy during Orion's tenure—even though it meant reducing production rates "during our prime money-making season"—because of her company's commitment to the environment: "as a company, it's just a value to us to minimize acid gas flaring events," Williams said. "It's something we track on our environmental scorecard, you know, and it's just a big no-no." In fact, the issue was so important to the company that, rather than being criticized for reducing short-term profits, Williams and her team were rewarded with the internal Chairman's Environmental Award for virtually eliminating flaring, among other improvements.

In St. Charles Parish's "responsible" petrochemical plants, then, corporate values for the environment itself and corporate commitments to being a "positive influence" in the community intertwined to support environmental initiatives that went beyond compliance with government regulations. Yet, once again, the rationale for environmental improvements came not only from the desire to reduce environmental damage or impacts on the community; rather, like other aspects of CSR, they were seen as integral to a plant's long-term economic success. John King explained what could happen if a facility came to be seen as a poor environmental performer or a risk to neighboring communities:

Well, it would be very hard to get permission to expand a business or a facility if you were demonstrably having impacts on people's health around you. So certainly expansions would be problematic and you would have to be probably directing a lot of investment into reducing your impacts, as well. Assuming that investment's available. So it's really about your viability as a business.

With only an oblique reference to the possibility of regulatory sanction, King made a business case for minimizing environmental impacts: significant impacts would compromise the facility's ability to expand, presumably jeopardizing its ability to become increasingly profitable, while simultaneously diverting revenue to solve environmental problems. In such a situation, he suggests, the company and its shareholders might be unwilling to make any further investment in the facility—making good environmental performance necessary for a plant's continued existence. Investment and reinvestment were the mechanisms cited by Randy Armstrong, as well, in his discussion of the need for ongoing environmental improvement. Regulation would not drive a plant out of business, Armstrong said; the real danger was that investors would "decide that you aren't of value in this area" and stop investing in renewing the plant's infrastructure, extending its life, and thus keeping it competitive. Maintaining the license to operate with the local community was essential to ensuring continued reinvestment, Armstrong suggested; if people became convinced that a facility's operations were environmentally unsound, investment would dry up, the plant would cease to be profitable, and it would eventually be shut down.

The logic of CSR, then, fostered not only environmental compliance but also environmental improvements at petrochemical plants by wrapping together community commitments, environmental responsibility, and business success. At the same time, the tight coupling of these three factors helped environmental managers navigate disputes over how good their performance ought to be. Citing the extent to which science had changed over the course of his career, and the likelihood that it would continue to do so, Randy Armstrong acknowledged that there were legitimate questions about what levels of air toxics, if any, were safe and that, with complete knowledge a near-impossibility, people perceived risks differently. Yet a plant like his could never drive its emissions down to zero, as some might want. So what it came down to, according to Armstrong, was, "How do you create a social environment to where, whatever either the real risk or the perceived risk was, that it was overweighed by the value that people had for us being here?"

In answer to his rhetorical question, Armstrong referred back to the kinds of programs that plant managers talked about being fundamental to their commitments to local communities, including helping to expand access to health care, improving educational opportunities, and so forth. What his account suggested was that, although environment was a core value for him and his company, environmental performance could never be adjudicated in absolute terms. At least within the logic of CSR, it was always a balancing act: environmental impacts, frequently of uncertain consequence, were to be assessed against (usually more tangible) social goods. The commensurability of environmental and community commitments, in turn, allowed environmental managers to resist demands for, for example, the complete elimination of toxic emissions: meeting such extreme environmental standards was not only seen as unrealistic; it was also not considered necessary for a facility to be of value to the community.

Responsible Company Men (and Women)

The February 2006 meeting of Valero's St. Charles Community Advisory Panel featured dinner and, unusually, a movie. Once the meeting was called to order and visitors were introduced, Ron Guillory played a short video that touted Valero Energy Corporation's achievements and accolades—and underscored various aspects of the refiner's commitment to social responsibility. The company was apparently proudest that it had been listed as number three on *Fortune*'s one hundred best places to work; the honor amounted to independent verification that the "Commitment to Our Employees" included in Valero's Commitment to Excellence was put into practice by the company, a point the film underscored by featuring both related honors and the company policies that helped make Valero a great place to work. The company's record of charitable giving, shown to compare favorably to that of other large corporations, was also mentioned in the video, as was the extent and reliability of its refining operations.

But the achievements in the video were not all that Valero representatives had to brag about to the CAP. In the announcements portion of the meeting that followed the viewing, Ellen Williams highlighted the St. Charles refinery's safety record: Valero employees had collectively logged three million work hours without any accidents or injuries that resulted in lost work time, and contractors at the site had completed another five million. They were especially proud of the latter, Williams explained, because accident rates are typically high among contractors; the five-million-hour milestone was testament to the good systems for communicating with and supervising contractors that the refinery had in place. The St. Charles refinery had also been designated a "Star" site in the Occupational Safety and Health Administration's (OSHA's) Voluntary Protection Program—a distinction, Williams said, reserved for facilities that went above and beyond to ensure the safety of their employees. Williams ended by touting her refinery's performance in internal awards competitions: they had won the Valero Chairman's top safety award for 2005 and come in second for the environmental award; the prior year, she reminded the group, the St. Charles refinery had won the Chairman's Environmental Award and come in second for the safety award. "Next year we're going to win them both," Williams promised. "I don't like second place."

Like the video that preceded them, Williams's announcements showed Valero's commitment to social responsibility by giving evidence of its Commitment to Safety—namely, eight million safe work hours and OSHA recognition. Yet she was simultaneously demonstrating her own effectiveness as the technical expert responsible for health, safety, and the environment: at her plant, the systems that she and her team had put into place had kept workers safe and earned recognition both internally and from regulators. Further, in performing her expertise—assuring CAP members, in effect, that the refinery was in the hands of people who were capable and motivated to run it well—Williams referred to a combination of her own personal characteristics and her company's status as a responsible refiner. With her assertion that she did not like second place, the petite, forthright woman with a slight drawl painted a picture of herself as ambitious, determined, and, ultimately, highly competent (for who competes for first place but those who know they can win?). It was her company's values, though, that framed the picture: the chairman's awards provided a context in which to express her determination and competence; they also allowed Williams to indicate that, as a technical professional, she valued excellence in the areas covered by the awards, safety and environmental performance.

In her performance of social responsibility at the CAP meeting, then, Williams's authority as an expert, especially the moral standing necessary to ensure her credibility with her audience,[27] was intertwined with the moral authority of her company, constructed through its CSR commitments. Top engineers and scientists at local sites of other large corporations similarly talked about their commitments in ways that elided their own moral standing with their companies' "responsible" status. For example, having noted that the top echelon of managers at petrochemical plants owned by large companies were typically promoted out of their positions, often into corporate management, within three to five years,[28] I asked Mitchell Mobley if the high rate of turnover made it difficult to maintain good relations with neighboring communities. I imagined that it might: would it not be hard to convince a resident of two or three (or more) decades that a cadre of managers who would only be there for two or three *years* really had the best interests of the community at heart? But Mobley assured me that it was not, because, while the individuals occupying top positions might change, the company's values did not:

> It would be a problem if there was lack of continuity in what a site's trying to do with a community. You know, if I came in and undid all the good that Mr. X before me did, then it'd be a problem. But that hasn't happened. . . . Because there's a thread in here that says our company desires to have this kind of influence, activity, presence in the community and [Mitchell], you know, you leave, and Linda comes in behind you and she's going to do the same thing.

Mobley's explanation, first, presumes that the individuals appointed to plant manager positions will share their company's commitments to nearby communities—reasonably enough, based on my experiences with St. Charles Parish plant managers and my understanding of what it must take to rise to such a rank in a company. But it also suggests that, as a result, the goodwill and credibility that a *company* has built up with a community will be extended to whatever new *individuals* come to occupy top management positions. In this construction, the personal integrity and values of each manager still matter: just as Ellen Williams's confession that she did not like second place served to bolster Valero's credibility by showing that the company's Commitment to Excellence was something that high-ranking managers made a personal value, one could easily imagine a plant manager who did not internalize his company's commitment to the community ultimately doing great damage to community relations. Yet the company's values were also integral to managers' ability to establish their personal integrity in the eyes of stakeholders: by extending a certain way of being in a community that the company has constructed as responsible, Mobley suggests, a new plant manager can enjoy credibility with community members even before they have had the opportunity to judge his or her moral character directly.[29]

Without the framework of a large company's commitments to social responsibility, the individuals at the helm of independently owned petrochemical facilities drew on more limited—and ultimately less powerful— repertoires for establishing their credibility. When Orion began operating the New Sarpy refinery, managers wanted to convince the community that it would not plague the community in the same way that it had as the Good Hope Refinery and as Transamerican. Called before the St. Charles Parish Council to give a "report on the Orion Refining Corporation" in July 2001,[30] Orion president and CEO Clark Johnson attempted to establish the trustworthiness of his operations primarily by referring to the personal characteristics of refinery managers. First, he stressed at length that they had nothing to do with the refinery's former ownership, underscoring the contrast by explaining that prior owner Jack Stanley had driven the refinery into bankruptcy, and that Orion had brought it out of bankruptcy. Johnson then went on describe the qualifications of people who were running the refinery as part of Orion:

> We've brought in senior operating people and technical people from other refineries all over the country to make this company an efficient, safe, reliable operation. Even going to our Board of Directors, all of our directors, again of the new company, Orion, are senior, retired executives from the refining industry. Conoco, Coke, Valero, Shell, all of them, and Lionel-Citgo refining, all five directors are senior, retired executives from the refining and therefore they have their, they have a high set of standards for themselves and for the company they're directors of. I personally have spent thirty-three years in the refining business. My first job, out of college, I went to Mississippi State University. My first job out of college was in Chalmette. I worked there, at the Tenneco refinery at the time, I worked there from 1968 to 1979 before I went to a hub office job working for Tenneco. But I did, I spent a lot of time in a refinery myself, have done everything you can do in a refinery.

In this statement, Johnson advanced the experience of senior Orion officials as evidence that the company can run the refinery well: he, the people on the company's board of directors, and the people in charge of day-to-day refinery operations had all spent decades running refineries, presumably safely and efficiently, and, Johnson suggested, would bring that experience to bear in running the Orion refinery. While the credibility of technical professionals arguably does rest in part on qualifications like these, the emphasis that Johnson placed on them distinguished his approach from that of managers with large, "responsible" companies. In the CAP meeting, Ellen Williams's promise that her refinery would have an exemplary environmental *and* safety record was backed up not by any explicit statement about how long or in how many places she had been a health, safety, and environment manager, but with a quick reference to an aspect of her character, namely, her aversion to coming in second. Johnson's testimony does include a character reference as well: he states that Orion's board members have high standards for themselves and the company. However, Williams's reference was able to be simultaneously charming and concrete, expressing her own value for safety and environmental excellence in the context of an institutionalized reward system for putting those values into practice; without the benefit of a CSR framework, Johnson's statement—which may have been thoroughly accurate—was generic and unconvincing.

A similar contrast was evident in the ways in which managers from multisited companies with CSR commitments and those from independent facilities talked about accidental releases. Both Orion and Valero engineers agreed that frequent flaring at the refinery, especially during 2000 and 2001, was a problem—but Orion's assurances to community members that they were fixing it lacked the weight of Valero's CSR-informed assurances. In his testimony to the Parish Council, Clark Johnson acknowledged that the refinery had had "excess flaring" as part of its start-up and expressed his regrets:

> I want you to understand that we don't like to be a nuisance to the community, and we really honestly were a nuisance to the community at that point in time. Uh, that's not the right thing to do, we want to be a good neighbor and that is a primary motivator in why we try to correct problems like that. But also if you can't run a, a coker unit is a key unit in a refinery, you need to keep it running reliably in order to be safe and efficient and to make money in a refinery. So point blank, we put all the effort that we could into the coker reliability, and the coker and other process units have increased in reliability tremendously since that point in time.

Johnson's sentiments, to a large extent, echoed those of managers at sites of large refiners: not being a nuisance to the community was a priority; the refinery wanted to be a good neighbor; and making money was unavoidably a goal. Yet the subtle differences are crucial: outside the context of a corporate commitment to social responsibility, not being a nuisance was something that Orion wanted—whereas, in Ellen Williams's account of Valero's decision to do something about the flaring, avoiding flaring and being a positive influence in the community were a matter of core values, fundamental enough to justify a shut-down that probably resulted in a short-term loss of revenue. In contrast, without the CSR apparatus that makes being a good neighbor integral to long-term profitability, Johnson cast flaring as a case where doing the right thing had to be balanced against making money.

Thus Ellen Williams may have been right that Orion's problems in New Sarpy had little to do with the moral character of top Orion officials; in both public presentations and interviews with me, they expressed their personal commitment and desire to operate safely, to avoid environmental impacts, and to treat the neighboring community well. And the drastic improvements in refinery performance that Williams and her colleagues were able to effect were certainly, as Williams suggested, due to Valero's value for safety and the environment and its orientation to long-term profitability—aspects of the large corporation's commitments to social responsibility that it supported with deep pockets and an extensive network of technical experts.[31] But the consequences of Valero's status as a responsible refiner were more than just material. The CSR framework that made the company a moral agent also helped its site-level managers to appear, at least, to be a "different cut of people." Where Orion officials could declare their credentials and assert their desire to do the right thing, managers at Valero sites (like those at sites of other responsible companies) could refer to their personal values in a way that invoked their corporations' commitments, lending weight and credibility to both sets of claims to moral authority.

Divided Authorities

In the aftermath of CCNS's campaign against Orion, then, New Sarpy residents became more willing to accept refinery managers' technical claims in part because of the sale of the refinery—not just to a different company, but to a different *kind* of company, one whose social commitments not only encouraged different environmental and community relations practices but also helped to ground site-level managers' technical authority. In Norco, however, no such transformation accompanied the end of CCN's campaign; indeed, Shell officials continued to tout the company's long history in the small town and its long-standing corporate commitment to social responsibility throughout the conflict. Moreover, Shell Norco managers ultimately resolved the community campaign—which had for many years charged that the chemical plant's operations sickened and endangered the community— by ceding to community activists' demands that they buy the homes of Diamond residents who wanted to relocate.

Such a move could have been painted as a tacit admission that the facility was harmful to community health and safety; it could have sabotaged Shell engineers and scientists' credibility with remaining residents of both the Diamond community and Norco as a whole. And yet it did not. Within a year of Shell's and CCN's joint announcement of the Diamond Options Program, public criticisms of Shell's performance seemed to have ceased, and remaining residents were, in fact, referring to the Shell-sponsored air monitoring program as evidence that there never had been an air quality problem in Norco. As in New Sarpy, the framework provided by CSR helped Shell managers to reclaim their technical authority—not, in this case, through new assertions of social responsibility but through strategic mobilization of the existing framework to articulate a rationale for buying out the Diamond community while remaining authoritative on technical matters. Specifically, corporate social commitments, by wrapping together profits, community programs, and environmental performance, extended the obligations of facility managers beyond their traditional areas of engineering and science expertise. As a result, Shell managers seeking to address CCN's campaign could identify social issues that they had failed to understand and had dealt with poorly; managers could admit to these shortcomings, moreover, because these were considered to be separate from their safe operations of the plant and entirely outside the area of their expertise as engineers.

The distinction between social and technical issues enabled by a framework for responsible business practices that included both—even in a technology-driven industry—was not only instrumental in allowing Shell managers to settle with CCN at little cost to their technical expertise but also central to the way in which facility managers across responsible petrochemical companies thought about maintaining good relations with neighboring communities. Taking for granted the soundness of their own approaches to environmental and health issues, they puzzled over how best to communicate the technical facts to community members and attributed potential community opposition largely to misunderstandings born of ineffective communication—again locating their own shortcomings firmly in a social realm distinct from their technical authority. Finally, the distinction allowed an opening for residents to participate in technical discussions of environmental issues, but only to the extent that their contributions could be understood as relating to social issues.

Expert Understandings and the Diamond Relocation

When they spoke to me six months after reaching a settlement with CCN, Shell Norco managers admitted that it was their actions—and inaction—that had made the relocation agreement necessary. But the problems they identified were not the plant's environmental performance or health consequences; rather, they represented the buyout as an acknowledgment that Shell had not fulfilled other aspects of its commitment to social responsibility in Norco.

For Wayne Pearce, the Diamond Options Program made amends for a Shell program that had unanticipated, negative effects on the community:

WP: The recognition there was that we'd done something when we started the work on building the greenbelt, which had a real impact in the community, one which we never intended, never foresaw.

GO: That was buying the first two streets to make the greenbelt?

WP: Yeah, and this is the agreement that we came to in Diamond that we referenced. So, this wasn't built on anything about emissions or anything like that anyway. It was more in that social end of the scale, you know, we recognized that the things that we had done on the first two streets, were the same, our intention was to make things better for people, making us more invisible in the way I described to you earlier on. But it had a real impact which came home to us when we really started listening, when we started to talk to the people in Diamond more.

What Pearce and his colleagues had realized, he went on to explain, was that buying the first two streets had fractured families and separated people from their caregivers—in direct opposition to Shell managers' desire to "make things better for people." In Pearce's account, then, the facility's mistake had not been poor environmental performance; on the contrary, their well-intentioned attempts to minimize the plant's impact on the local community had backfired.

David Brignac similarly downplayed environmental factors in his account of the agreement: "Well, yeah, our performance hasn't been perfect, but the deeper underlying issue seems to be some resentment that [Diamond residents] haven't shared in" the high-paying jobs and other benefits of being host to a large manufacturing facility. Brignac's analysis of the situation referred back to the idea, central to the CSR framework, that real and perceived environmental risks could and should be balanced out by corporate contributions to the local economy and quality of life. At the same time, though, Brignac acknowledged that the racial discrimination and segregation characteristic of the American South until (in his telling) the late 1960s had played a large role in excluding Diamond residents from the benefits of Shell and was an even deeper source of resentment:

So that's some underlying issues there as well, it goes back to racial discrimination practices or whatever in society at large. But probably within Shell too, I mean Shell is part of society, but even though times have changed, they changed decades ago really, because Shell for a long time now has publicly [said] and demonstrated that we're an equal opportunity company, we don't discriminate on the basis of religion, sex or whatever, but you still have those old feelings because the people who lived through those bad times are still there.

Here again, Brignac suggests that the buyout was precipitated by Shell's failure to live up fully to its social commitments: the company's current values for diversity and nondiscrimination had not been part of Shell Norco's early interactions with African American neighbors, and its complicity with racist practices widespread in the mid-twentieth century had fostered what Brignac felt was understandable resentment against the facility. Their resentment, he thought, made Diamond residents more critical of Shell's sound but imperfect environmental performance than they would have been if they had not been excluded from Shell's jobs and community programs.

When talking to me about their mistakes in Diamond, Shell managers attributed them to their own lack of understanding of social dynamics in Norco—a kind of understanding, several of them noted, outside their usual purview as engineers. Describing the buyout of the second two streets in Norco as "the only option," despite the fact that he "felt very bad when [it] was announced,"[32] Randy Armstrong said that it was Shell's failure to understand community needs that made the move necessary: "We sort of reached an impasse. And that was just created by not responding to those needs for a long time. Maybe reality was, not even being aware of them." Armstrong went on to illustrate Shell's lack of awareness by describing a study undertaken by the EPA as part of their Common Sense Initiative: "What came out of that was, 'oh, by the way, you've got five communities [within Norco] and you're only talking to one of them.'" The finding was "shocking," according to Armstrong; it was "the very first time" he and his colleagues had realized that there were such divisions. Like those of other Shell managers, Armstrong's account of the buyout blamed his organization's failure to meet its self-imposed social responsibilities but saw the failure as stemming from fundamental gaps in what they knew about the community: because they had not recognized the heterogeneity in Norco, Armstrong suggested, Shell's outreach and community programs only served a fraction of residents, which created disparities, resentments, and heightened sensitivities to environmental risks that, ultimately, could only be resolved with the relocation agreement.

While Shell managers acknowledged shortcomings in their understanding of the community, they contrasted the insights necessary to fulfill social responsibilities to their engineering expertise. Pondering the challenge of how to do good in a town that they now recognized as made up of several different communities with diverse needs, Randy Armstrong remarked that "it's been a real interesting experience for this hard-core engineer to try to practice social issues." David Brignac similarly described how he, as manager of the Good Neighbor Initiative, had had to learn that not everyone thought about things the same way he had been trained to as an engineer. He gave the example of Shell's 2001 decision to purchase the overgrown, uninhabited tract of land that separated the Diamond neighborhood from the rest of Norco, in order to get the riverfront property attached to the tract—a decision that sparked outrage among residents who were calling for relocation:

> The family said, we're not going to sell you this riverfront you want unless you buy this piece. If you looked at the price, the price of the riverfront was very expensive, the wooded area was not very expensive, but they wanted it off their hands, because the parish was putting pressure on them to clean it up. And so, it's uninhabited, so we're not buying anybody's house, right? And then, so, the company bought that . . . and that piece, and then the Diamond residents erupted, like "Oh, you bought these trees and this woods, and you won't buy us." And the logical argument was, "We paid for this whole big wooded area what we'd pay for one house. So you're talking apples and oranges in terms of logic. So why are you even upset?"

The unanticipated reaction from the community illustrated a key lesson that, Brignac said, he had had to learn in his job: "You tend to think logically as a technical person, but you understand that issues in the community are not logical, they tend to have to do with feelings, and someone . . . years ago and resentment and. . . . So you learn that it's not scientific, it's behavioral." In order to make sense of the conflict between Shell and its neighbors, then, Brignac drew a sharp distinction between the social and the technical, the emotional and logical. His point in doing so, moreover, was not to deride community members—indeed, he offered that "what you'll find is that it gets back to who are the weird ones? It's not the people, it's probably the engineers"—but to underscore, as Armstrong did, how far removed the social arena was from his own areas of expertise.

Shared among top Shell Norco managers, then, was a story of the Diamond buyout that admitted significant failings on the part of the company, but that distanced those failings from areas in which the engineers in charge of the plant claimed expertise. At least in one segment of Norco, Shell had not been the positive influence it was committed to being because, the story went, engineer-managers had not understood the diversity of Norco and the needs of the Diamond community in particular. These gaps in social knowledge, combined with reasoning processes characteristic of engineers, had led them to decisions that contributed to, rather than ameliorated, the difficulties and resentments of Diamond residents. In this story, Shell Norco managers start out ignorant of important community dynamics, and their growing understanding leads them to craft a buyout agreement with CCN that redresses the harms they have done to that segment of the community. But there is no comparable trajectory for engineer-managers' understandings of their plant's effects on environmental quality or community health in Diamond. Rather, Shell officials' account sidelines CCN's grievances about dangerous operating units, poor air quality, and exposure-related illnesses[33] and takes for granted that the environmental and health effects of Shell's emissions were both well understood and well controlled by plant personnel all along.

In this case, then, the social commitments of the responsible petrochemical company created space for facility managers to resolve an environmental justice conflict without ceding any of their technical authority. That is, both Pearce and Brignac were explicit about being unwilling to entertain allegations that Diamond residents were harmed or endangered by plant operations as a justification for the buyout; if the campaign had had to be fought out purely on the basis of residents' deep concern for community health and safety, it is likely that Shell and CCN would still be at odds—or that CCN's campaign would have met a fate similar to CCNS's. But Shell Norco's status as a facility of a socially responsible global corporation created other possibilities. Because Shell's CSR framework obligated Norco managers to the well-being of the community broadly, the same logic that led them to invest in adult education programs allowed them, eventually, to find a justification for offering relocation to the whole of the Diamond community—a justification that, crucially, depended on admitting shortcomings only in areas where engineers, by virtue of their technical specialization and particular ways of thinking, might be expected to make mistakes.[34]

The Challenge of Communication

In Norco, by extending the domain of petrochemical facility managers' obligations beyond the traditional realms of scientific and engineering expertise, Shell's CSR commitments separated out a "social" space where managers could take responsibility for community grievances in ways that deflected criticisms of plant performance and of industry experts' claims about its effects on community health. But the same dynamic also operated more broadly. In their routine efforts to maintain good relations with neighboring communities, facility managers for responsible petrochemical companies took for granted their mastery of the technical aspects of plant performance and saw potential for community opposition to emerge from failures in the social realm. Specifically, imagining that opposition was likely to be fueled by misunderstandings on the part of residents about the real threats posed by facilities, the engineers and scientists at the helm of not only Shell but also Valero and Motiva puzzled in particular over how best to communicate technical content to the community. In the process, communication became a general mechanism for acknowledging obligations to the community while refusing criticisms of environmental performance; that is, opposition was usually acknowledged to stem from failures by plant officials—specifically, their failure to help neighbors make sense the operations and impacts of a facility—but their shortcomings were again considered to belong to a realm in which they, as engineers and scientists, were understandably inexpert.

My 2006 interview with Ellen Williams took place the day after a Valero CAP meeting that we had both attended. Comparing notes on the meeting, I marveled at the lack of interest in environmental and health issues that resident CAP members seemed to express in setting the panel's agenda for the year, and I was surprised by Williams's response. She said that, although she was relieved that residents, for the first time, had not "chewed on her" about the beneficial environmental projects (BEP) that Valero had inherited from Orion, she was disappointed that they did not take a more active interest in environmental issues, even to the point of challenging her on plant performance.

> EW: Last night, once we got past the BEP stuff and I realized Miss Ida wasn't going to fuss at me again, I found [the CAP meetings], I find them boring, almost. This one, because it's not, sometimes some of the hard questions and the demands for explanations is more interesting anyway, especially when you know you have a good story to tell, than sitting there and saying how great we are. Don't get me wrong, I'm glad they like us. I think we do a good job, you know, we're sincere. But I don't mind the tough questions, you know, and I want to show off more about how good we are.
>
> GO: What kind of question is a tough question?
>
> EW: Oh, well, you know, just like explaining . . . we used to go over our [environmental] score card . . . explaining why we had this event and what we're doing about it.

Rather than hoping community members would be uncritical, then, Williams was actually eager for "hard questions." Such questions were not threatening, her comments indicated, because she was confident that she and her team did a good job. And they were welcome because they offered Williams an opportunity to demonstrate—rather than just assert—how well they were doing through explanations of some of the details of the plant's operations.

Williams was not the only high-ranking manager to want to engage "hard questions" from community members. Mitchell Mobley contrasted a CAP that he considered "a bland situation" with one that he regarded as especially useful, citing one member of the latter as exemplary: "She asks good, pointed questions about environmental performance, about health hazards, she, we know we can count on her to be very business-like, to ask tough questions." Like Williams, Mobley was unworried about his ability to give satisfactory answers to such questions; rather, he saw them as occasions to help community members understand how plants operate. Asked to give an example of a tough question, Mobley cited a recent incident in which plant managers of facilities significantly smaller than his boasted of the number of weeks and months they had gone without any environmental exceedances. When Mobley's turn came,

> I stood up and I said, our last environmental exceedance was yesterday, and in the course of the year, we've had forty of them, and we're working awfully hard to change that. And [a CAP member] stopped me and she said, "Okay, [Mitchell], well, tell us, you know, why it is the way it is and what you're doing about it." . . . I don't have any problems talking to her about the areas where we have problems or what we're doing about it. And I don't compare ourselves to those other industries. The complexity, the kind of stuff we have floating around in here that they don't have floating around in those plants. It's just, it's a different world. . . . But the community doesn't see that, you know, it's industry. And I know what courses through their minds, well, gee, if Oxychem can have one every 130 days, why does [Mitchell] have to have one a week, or two a month, or whatever.

Even though the incident highlighted an area where plant performance left something to be desired, Mobley's telling suggested that he did not see it as potentially shaking community members' confidence in the overall safety of his plant. On the contrary, he talked about how it gave him a chance to explain a subtle point often missed by community members: not all industrial facilities are alike, and it did not necessarily make sense to judge them all by the same standards.

In their confident answers to tough questions, facility managers take for granted that the explanations they are providing represent the one right way of looking at the issue in question. As a result, they see themselves as educating community members on important facts and, in many cases, countering misinformation. Mobley's attempt to explain how his plant differed from other local chemical facilities, supplanting residents' view of industry as homogeneous, framed environmental exceedances as a function of the complexity of industrial processes—ignoring equally reasonable ways of making sense of exceedances by, for example, examining their size, composition, and likely effects without reference to their source. Similarly, Ellen Williams explained how she hoped that resident members of the CAP would come to understand flaring better:

> The way I think CAPs ought to be used is you educate the CAP around your issues and then they're kind of like your goodwill representatives in the community that can say, you know, "Yeah, I know about that flare when it goes off, and really it's a safety device." And those kind of things, "and the company doesn't want to put things in a flare any more than we want to see stuff in the flare. It's money, it's their product, so they think that's bad, or whatever, and it provides a safety aspect to running the plant." And just things like that.

Imagining CAP members correcting the misconceptions of other residents—explaining what a flare "really" is—Williams presented as uncontroversially true industry engineers' understanding of flaring, which, while reasonable in itself, is nonetheless widely contested by environmental activists who argue that the cumulative effects of flaring on fenceline communities are much greater than calling it simply a "safety device" would imply.

Facility managers, then, were interested in tough questions not as a way to open up discussion of petrochemical industry impacts from diverse perspectives but as an opportunity to school community members on the nuances of plant operations as engineers understand them. Moreover, they imagined that educating residents would minimize the likelihood of community opposition. In Ellen Williams's example, community members who knew about flaring would be able to interpret the sight of a flare for their neighbors, casting it as evidence that the company was operating safely and responsibly—rather than thoughtlessly polluting. Mitchell Mobley, too, imagined that his explanations to CAP members could help mitigate potential opposition: "If it gets to the point where people are taking public stances against our business, they can be an educated faction, if you would, that can add some reason if reason doesn't exist." Mobley's vision, notably, implies that "reason" belongs to those who have been educated in engineers' framings of environmental and safety issues and hints that public opposition to the petrochemical industry is likely to be characterized by a lack of reason.

In this view, not having the opportunity to answer tough questions with reasoned engineering explanations was considered dangerous to a company seeking to maintain good relations with its neighbors. Describing the difficulty of getting health and emissions information out to the whole of a community, Wayne Pearce noted, "It's happening today, I'm sure, there's all sorts of things going on which create questions and concerns in people's minds, and unless we're accessible and listening, those things can grow over time and become something much bigger than they needed to have been." Environmental campaigns like that in Diamond, in Pearce's view, were built on concerns and misunderstandings that, left unaddressed, festered; listening for tough questions was thus central to avoiding opposition. Ellen Williams likewise suggested that the campaign in New Sarpy was attributable to poor communication. Had Orion been better at communicating, she said, "They probably would have gotten their residents together and said, 'Okay, we're going to start this thing up, and you're going to see some flaring, you're going to hear some noises, and here's what it . . .' You know, they missed an opportunity." Here again, Williams expressed the belief that communicating engineers' ways of looking at plant operations—in this case, letting the community know what the flaring during start-up meant and why it was necessary—would lessen or even eliminate residents' opposition.[35]

With good community relations seen to rest in large part on their ability to educate residents about the technical issues related to petrochemical plants' operations and effects, facility managers regarded effective communications as one of their primary challenges. In interviews with me, they described the challenge and speculated about how they might improve their efforts. Some focused on the difficulty of relaying technical information to an audience without technical training. For Wayne Pearce, conversations about the effects of toxic exposures on health presented a particular dilemma: in trying to answer the question of whether the health of people in Norco was harmed by the emissions from industry, he said,

> You get into, let me get the right words, morbidity studies, and incidence of cancer studies, then you get into statistical analysis, and very soon, you're beginning to lose the message with the community at large. Which I count myself in that one, as well, you know, unless I have a lot of time to sit down and get people to explain in a lot of detail what they're talking

about, the debates between the scientists go off on a track which also loses people, because it's not in layman's terms, in readily understood terms, so then both sides slip back to explain things in layman's terms, and they lose the main point of the discussion as well.

Both highly technical explanations and those easily appreciated by nonscientists, Pearce suggested, ran the risk of not being very helpful as a way of addressing community concerns: the technical ones because even someone like him—a chemical engineer rather than an epidemiologist or statistician—could get mired in simply trying to understand the science and lose the larger issue, and the simplified ones because they missed important technical nuances. The goal, he said, would have to be "making it simpler without rejecting the complexities," and he cited regulatory standards as useful in conveying a simple message that was nonetheless solidly grounded in science.

Beyond puzzling over the difficulties of presenting complex scientific and technical concepts, high-ranking managers pondered the structures that would best foster communication. All agreed that CAPs were useful, but limited, and contemplated ways in which they could be improved. Ellen Williams's disappointment with the previous night's CAP meeting prompted her to think aloud in our interview about whether the CAP might be better served by an in-house facilitator rather than a third-party one. And Mitchell Mobley mused that rotating people off the "bland" CAP more systematically could possibly make the group more dynamic and allow industry representatives to communicate with a larger portion of the community. In thinking through both communication structures and the particular challenges of technical information, facility managers acknowledged communications as something they were accountable for—communicating effectively was central to being a positive presence in the community, which was central to their values as managers for responsible companies. At the same time, it was not something that they claimed to know how to do effectively, the way they knew how to run their facilities well; instead, it was an ongoing challenge outside their areas of expertise.

Like Shell's explanations for the Diamond buyout, petrochemical companies' model of the role that communication played in community relations attributed opposition to factors that, while within the purview of the responsible petrochemical facility, were not central to the expertise of the engineers and scientists who run facilities. Admitting to poor communication thus became a way for plant managers to take responsibility for poor community relations without owning up to shortcomings in their technical performance.

In the year following their settlement with CCN, for example, Shell officials indicated that the most important lesson they had learned from the experience was that they needed to be communicating more with the community—not only conveying information to them but also, in Pearce's words, being "accessible and listening." No one hinted that their environmental record might also need to improve; that, they would have said, they had been doing a good job of all along.

The use of communication as a device to allow facility managers to concede some shortcomings while rejecting allegations of technical incompetence or irresponsibility was even more striking in Orion officials' accounts of community relations in New Sarpy. In his testimony to the St. Charles Parish council, Orion president Clark Johnson took pains to establish that he and his staff—whose credentials he had asserted—were running the refinery responsibly. He described the excessive flaring that was, he admitted, a nuisance to neighbors as actually just part of "typical start-up issues in a refinery." And, in his account of the June 2001 tank fire, Johnson described in detail how refinery employees had responded proactively to what was, in his telling, a natural disaster: in a routine check, operators had noticed gasoline on the roof of the tank as a result of torrential rain; staff had immediately started draining the tank, while trying to get the gasoline-and-water mix off its roof without violating environmental regulations; and managers had anticipated the nightmare scenario of a lightning strike and begun to put emergency responders in place even before a bolt ignited the remaining gasoline. They had, in short, done everything they could do and had extinguished the fire, which ultimately burned for over thirteen hours, as quickly as possible. The one place where Johnson was willing to acknowledge problems with his staff's handling of the event—and their operations of the refinery more generally—was in the area of communication. Admitting that some neighbors were upset with Orion, he outlined some fundamental improvements:

> We need to talk more to the neighbors, we need to find out what their concerns are, we need to find better ways to communicate to the neighbors, so we are setting up some agenda to do that right now, to meet, to discuss, to hear more from the neighbors about what their concerns were.

Even after a period of frequent flaring and a massive fire, then, Johnson's proposed remedy for an angry community did not include any mention of improved emissions controls, safety systems, or plant reliability—only improved communication. Moreover, his brief description of what it meant to communicate better echoed that of managers from other companies, suggesting that he was hoping to convey Orion's way of understanding environmental risks and answer community concerns by reframing them in terms favored by engineers.

Framing Community Questions

As a mechanism for avoiding and accounting for community opposition, communication mobilized the same divide between the social and the technical that Shell's description of the Norco buyout did, and it rejected the notion that residents could legitimately object to a nearby plant on the basis of its performance or impacts. If community members really understood the workings of the plant, the logic went, they would know that the engineers running it were doing so as safely and responsibly as possible. But not all community concerns were ruled illegitimate or ill informed in this framework; on the contrary, the social-technical divide enabled managers at petrochemical plants to engage with community members even around environmental issues—as long as residents' contributions could be framed as social ones.

The utility of the distinction was especially apparent in the "Air Monitoring...Norco" program, which Shell and Motiva undertook as a beneficial environmental project mandated by a settlement with environmental regulators. Shell officials clearly hoped that it would conclusively counter allegations of poor air quality made by bucket-wielding CCN members and were, as a consequence, concerned with making its results credible to community members. Residents were thus invited into the planning process from the beginning. The way that process was structured, though, reflected the same separation of social and technical issues that characterized responsible companies' approach to communications: work was divided up between the "Technical Team," charged with deciding how to do the monitoring, and the "Communications Team," charged with figuring out how best to get the results out to the community. The technical team consisted of technical people: high-level environmental health and safety staff from Shell and Motiva, a specialist from one of Shell's corporate R&D labs, the person in charge of Louisiana Department of Environmental Quality's ambient air monitoring program, and a contractor from URS, the environmental consulting firm that would conduct the actual monitoring. The residents invited to participate in the process were all assigned to the "Communications Team," along with a public health researcher from Tulane University and a junior, nonengineer member of Shell's Good Neighbor Initiative staff.

The neat division was soon challenged, however: the residents on the communications team wanted to participate in technical team meetings, and Shell and Motiva officials readily agreed to let them. Moreover, they were surprised at residents' contributions to discussions about how the collection of air quality data should proceed—or, rather, *where* it should occur. According to John King, one of the original members of the technical team, residents' interest in monitoring site locations contrasted sharply with what he thought of as the interesting issues:

> The thing that came out over and over again as important for people living in the community was the locations, where the monitors were. If I just sat down with an academic or an engineer's point of view, I might have guessed that they would be interested in other things, around sampling technology, or were the canisters mirror-glazed, double-lined or glass-lined, you know, or . . . how long the samples were analyzed for or how frequent they were or whether a chemical was or wasn't on the list. Those are all important issues, and I think we had people involved in the process who were academic experts who were tuned into those. But the community added an element of what's the reality of the neighborhoods in Norco and where people perceive that they live and what they perceive their airshed to be.

That residents were interested in the location of monitors over other details of sample collection and analysis came as a particular surprise to members of the technical team because they believed that it would not matter where air samples were collected: they had good reason to expect that air toxics concentrations would be the same across the small town and that, as a result, a single air monitor placed just about anywhere would be able to give a good picture of air quality. But through community members' participation in technical team meetings, they came to understand that community members would not be satisfied with a single monitoring site and so developed a two-phase plan that included sampling at six different sites during the first phase. Karl Loos, a Ph.D. air quality consultant from Shell Global Solutions (also white and in his fifties), explained why they were willing to do so:

> Initially there was a big thing of, "Well, you're going to monitor air in Norco, well, *where* are you going to monitor air in Norco? I live on this street, I live on this street over here, how do I know that the air is the same?" It was a big issue. You know, what a technical person would do is say, well, Norco is a well-mixed community, it's relatively small

geographically, the wind field varies, and dispersion modeling calculations have shown that the air is uniform, relatively uniformly mixed in Norco. So therefore just one site is necessary. If we would have said that, people would have felt we were trying to put something over on them, I think. Because there was a real discussion, you know, "Well, I'd like to know what the air is where I live." . . . And while we really believed what we said was true, it's always good to test those assumptions periodically. So that's why the phase 1 monitoring program.

Like King, Loos noted the significant differences between the way technical team engineers were inclined to frame the problem of characterizing air quality in Norco and the way community members did so. For Loos, the issue that the differences created was one of credibility: if engineers ignored community members' problem framing, they ran the risk of the community not believing their claims about Norco's air quality. But he also admitted a very small sense in which residents' concern with monitoring sites was not just a problem of credibility and trust—that is, a social issue—by allowing that it gave engineers the opportunity to test their assumptions about the uniformity of the air in Norco.

With the release of Phase 1 results, however, community concerns were relegated entirely to the social sphere. The data from four of the six monitors showed good agreement, and the remaining two monitors showed higher levels of chemicals that Shell engineers were ultimately able to track to, in one case, "substandard" seals on a set of storage tanks and, in the other, ground-level emissions from a different industrial facility. Reports on the results thus concluded that "Norco's air is fairly uniform," and engineers saw their models of Norco's air as having been confirmed because, as Loos put it, the data showed that air pollution was distributed uniformly "with a few exceptions, and every one of those exceptions could be explained very cogently." While technical team engineers emerged from the experience with renewed confidence in their science, they nonetheless told me that they were happy to have gone through the lengthy process of involving community members and incorporating their concerns. In addition to making the study credible, as Loos suggested, community participation in the technical team gave plant engineers the opportunity to educate a small group of residents; King's description of the process—"people kind of came up a knowledge curve with us as we talked about the issues and learned more about how air monitoring was done and what some of the tradeoffs were"—suggested an accomplishment similar to that hoped for by industry participants in CAPs.

Shell's monitoring program was remarkable in that, by translating residents' concern with variations in air quality between neighborhoods into a plan that included multiple monitors instead of the engineers' one, it allowed community members to influence the production of scientific knowledge—despite an initial structure that assumed they would not. Yet residents' participation in the technical team was not recognized as a technical contribution. That is, instead of being seen as "exceptions" to the otherwise uniform air that engineers had predicted, the noticeably higher concentrations of chemicals at two of six monitoring sites could have been interpreted as demonstrating the highly localized significance of so-called fugitive emissions—the chemicals that escape, often at ground level, from leaky valves and seals as opposed to those released from high stacks—to ambient air quality; such an interpretation might have pointed to the need for some sort of monitoring grid as the best way to understand air quality, rather than justifying a gradual reduction in the number of monitoring sites. Had the monitoring program's results been interpreted in this way, residents' contributions to scientific understandings of the impact of petrochemical emissions on air quality would have been obvious. But in the interpretations offered by Shell, community members' involvement was terribly important to the credibility of the data and contiguous with the company's efforts to communicate—and thoroughly irrelevant to scientific understandings.

Notably, both residents' (arguably) technical contributions and their framing as social ones were enabled by Shell's corporate values. The responsible company's desire to maintain good community relations, to address community concerns, and to be credible with community members—rather than just to operate in keeping with what engineers consider best practices—prompted them to design an air monitoring program that answered residents' questions as well as their own. Yet the concern for issues beyond technical effectiveness or efficiency also created a space in engineering practice where residents' questions could be understood as something other than a contribution to knowledge production, and thus be taken up without jeopardizing engineers' claims to expertise.

Conclusion

During my fieldwork, I became friends with Donna,* a white single mother only a few years older than I, who had recently come from a public sector job in a neighboring state to work on Shell's Good Neighbor Initiative. Over a glass of wine on her patio one spring evening, Donna told me that many of her former coworkers considered Shell a big, evil company and confessed that, especially after all the fuss in Diamond, she honestly wondered if there was truth in their view. So she went in with her eyes open, and after a year, she really could not find any evidence that her colleagues and the company were interested in anything other than doing the right thing. She supposed that there was still the possibility that she did not have the whole picture, but "if they're evil," she said to me, "they're sure hiding it well."

Donna's observation echoed my experience with Shell and its peers in St. Charles Parish. I never had any reason to believe that the facility managers whom I interviewed were dissembling; indeed, they appeared to act with integrity in every context in which I observed them. A few, in fact, went out of their way to make sure that my questions about their operations were fully answered, and all spoke thoughtfully and compassionately about the communities neighboring their facilities. When, in teaching engineering ethics at the University of Virginia some years later, I asked my students to put themselves in the shoes of David Brignac—a chemical engineer facing tough ethical issues as manager of the Good Neighbor Initiative—I did so not because I thought they could imagine better resolutions to the situation in Diamond than Brignac had but because I wanted to give them a real-life example of someone who had acted well in response to a real dilemma.

Yet it was also clear that Brignac and his colleagues acted with integrity within the context of a particular framework—that of corporate social responsibility. Taking profitability as a central value, accepting responsibility for the well-being of neighboring communities, and understanding environmental risk and social benefit as fungible but technical and social competencies as separable, the CSR framework is just one of several possible ethical stances. Choosing it over, say, an ethical framework that emphasized justice[36] had significant consequences for petrochemical industry scientists' and engineers' ability to recover and maintain their authority as experts in the face of community opposition.

Other scholars have critiqued CSR for imposing particular notions of "community" and constructing community needs in a manner that (inappropriately) focuses on individual responsibility.[37] Their analyses are borne out in New Sarpy, where Valero officials praised Orion's involvement in constituting the St. Charles Terrace Neighborhood Association to *really* represent the community, and Norco, where residents' environmental justice concerns were framed in community education programs as being about equitable access to the good jobs offered by the plant. But to these critiques I would add that, in the case of environmentally hazardous industrial facilities, the CSR framework constructs the "social," as an aspect of technical practice, in ways that are equally disabling for community-based efforts to secure environmental quality and social justice. By wrapping together profits, community obligation, and environmental responsibility as all part of the same economic (or engineering) rationality, the CSR framework contains the suggestion that the health of communities and the environment can be traded off against economic benefits. The fact that responsible companies appear to operate facilities at a much lower impact on the environment, as illustrated by the New Sarpy refinery's transition from Orion to Valero, does not change the injustice inherent in the idea that money—whether in the form of jobs or funding for community programs—can compensate people for assuming risks to their health and safety.[38]

More consequentially, by creating a social space in a technological business, the logic of CSR offers opportunities for engineers and scientists to ignore, deflect, and misrecognize legitimate community criticisms of technical people's accepted ways of thinking about environmental impact and community health. By casting complaints about flaring, accidents, and air quality as either misunderstandings of technical issues or veiled grievances about a facility's social performance, facility managers protect their own authority over environmental issues—and disable attempts by residents and their environmentalist allies to participate in the framing of environmental health questions and the production of knowledge about them.

Corporate social responsibility may represent a significant improvement over the ethical stances adopted by engineers prior to the environmental and human rights movements that prompted its development, and it certainly seems to produce better outcomes for communities than ethical positions that are not grounded in a framework that sees community, environment, and profitability as interdependent. Science and technology studies (STS) scholars and educators, invested as we are in treating knowledge production and technological development as inescapably social processes, suggest that current ethical stances might be improved still further by refusing simple distinctions between social and technical and imagining so-called laypeople not just as recipients of knowledge and technical solutions but as potential collaborators in defining and approaching technical problems.[39] But analyzing how the social-technical divide serves the interests of responsible petrochemical companies seeking to avoid, or recover from, community opposition, sheds light on the real obstacles to such an STS-informed improvements: the petrochemical industry has much to gain from being socially responsible and technically unassailable.

The neoliberal logic of responsibilization that grants moral authority to corporations, and gives rise to CSR programs, enables large petrochemical companies to stake out such a position. From that position, in turn, plant-level engineers and scientists can construct themselves as responsible experts, sensitive to community needs and even humble about their understandings thereof, but authoritative on technical issues; from that position they can neutralize activists' critiques of their environmental practices and knowledge claims by reframing them as social, rather than technical, challenges.

6

Passive Revolution and Resistance

In 2007, I was invited back to New Orleans to celebrate the tenth anniversary of the Louisiana Bucket Brigade (LABB) and participate in a weekend-long retreat to establish a strategic vision for the organization's future. On a not-yet-steamy Saturday morning in July, LABB staff, board members, volunteers, and community partners congregated in a restored Creole villa in the Treme, home to the New Orleans African American Museum. We began by reflecting on LABB's major activities and achievements, and the "learnings" that had come from them. Participants recalled working with Greenpeace on campaigns in southwestern Louisiana that included nonviolent resistance and giant banners hung illegally from highway overpasses—tactics that, founding director Anne Rolfes noted with some regret, had faded away from LABB's work. The relocation of the Diamond community was of course also added to the giant timeline we constructed on one of the few walls in the villa not hung with artwork. So, too, were LABB's more recent activities in Chalmette: there, together with St. Bernard Citizens for Environmental Quality and an especially skilled full-time volunteer, the organization had not only used a sophisticated open-path air monitor to document a violation of the EPA's 24-hour sulfur dioxide standard; it had also managed to force the neighboring Exxon oil refinery to pay for ongoing, real-time air toxics monitoring at three locations near its fenceline.

But New Sarpy was left off the timeline—even though the campaign there had been LABB's primary focus throughout the second half of 2002. When it became clear that none of the other retreat participants was going to mention it as a significant event in LABB's history, I ventured to suggest "First Clean Air Act lawsuit in New Sarpy" and "New Sarpy settlement" as milestones that should be included. Always generous in validating people's contributions, Rolfes immediately said that it was good that I had brought up New Sarpy, that we should be talking about what we had learned from it. The handful of us who had been involved with LABB at the time—including Karla Raettig, a white lawyer not much older than I, who had been part of the Tulane Environmental Law Clinic team that represented Concerned Citizens of New Sarpy (CCNS) in the suit, and Iris Carter, a former Diamond resident in her fifties, as well as Rolfes and I—talked through the incident, explaining to those who were not there how frustrating and disappointing it had been to watch New Sarpy residents end up with such a raw deal. The professional facilitator running the meeting questioned our assessment of the situation: even if *we* wanted something better for the community, could it not still be a success, as long as *they* were genuinely happy with the outcome? My colleagues granted her point, but refused to accept the suggestion that the campaign in New Sarpy had been a success. The community had never really coalesced around campaign goals, Rolfes said—and the need to clearly identify community goals was recorded as a "learning." Raettig pointed to Orion's behavior as another factor that made it hard to consider CCNS's campaign a success: what she had learned from the experience, she said, was how manipulative companies could be and how far they would go to get their way.

I cannot disagree with Rolfes's and Raettig's views: indeed, they identify critical aspects of New Sarpy residents' encounters with Orion that are underemphasized in the analysis that I have presented here. CCNS members—even the small group of leaders—did not have a unified vision of what they cared most about or what they wanted to accomplish, and when Orion began to apply pressure to the group, the fault lines became decisive in the campaign. And Orion's tactics—offering cash to low-income families just in time for the holidays; incorporating a community group to supersede the existing, if loosely organized, CCNS as the voice of the community; and stalling legal proceedings to push for a settlement before the lawsuit could reach the discovery phase, when additional information about the refinery's environmental record would have been revealed—were blatant, manipulative exercises of corporate power.

As significant as these factors are, the assessments offered at the LABB retreat underestimate what community campaigns like that in New Sarpy are up against. Deeply ingrained understandings of what it means to be a responsible person and a good citizen—shaped by our nation's liberal traditions and its neoliberal practices—make the costs of environmental justice campaigns exceptionally high for communities, perhaps especially so for the low-income and/or minority communities that are most likely to need to resort to collective action in order to protect their health and environment. Asserting that they were stuck in a toxic environment not of their own choosing jeopardized New Sarpy residents' status—and sense of themselves—as enterprising individuals, capable of and committed to pursuing a good life for themselves and their families. By declaring their neighborhood toxic, they risked tarnishing the image of the small town in the eyes of outsiders on whom future investment depended. And in refusing to engage with Orion representatives without the support of nonprofit allies who could call into question the presumption of equality structured into industry-sponsored talks, they were all too easily dismissed as simply impossible to reason with.

Notions of responsible choice, entrepreneurship, and reasoned, egalitarian discussion thus create significant ambivalences for residents of fenceline communities. Collective action may be the only choice for a community wanting to improve its environment and mitigate the health effects of industrial pollution; however, resting on the premise that, acting alone, even enterprising individuals cannot persuade a powerful company to change its practices, it is far from a natural or obvious choice for many, if not most, communities.[1] This endemic ambivalence certainly made it possible for Orion to manipulate New Sarpy residents—and the more explicit, coherent campaign goals that Rolfes would have liked CCNS to have had may or may not have been enough to overcome deep-seated biases against collective action.

But the same understandings of personhood and responsible action also make it possible in many circumstances for petrochemical companies to secure the acquiescence and even support of fenceline communities without resorting to Orion's brand of heavy-handed tactics. Large, multisited companies with explicit commitments to social responsibility—companies like Valero and Shell—have acknowledged and now work to address many of the justice issues raised by grassroots environmental campaigns, albeit within the particular framework of corporate social responsibility. Aware of the argument that those most affected by chemical plant emissions benefit the least, they fund educational initiatives designed to give locals the skills to work as plant operators, and they work with residents to identify other community needs to which they can contribute. Sensitive to charges of racism, they emphasize the diversity of their workforce and even, in rare cases like that of Shell Norco, adopt programs to benefit victims of former, racist practices. And, interpreting communities' desire to have a say in decisions that affect their health and environment as the need for open communications, they make themselves accessible through Community Advisory Panels and other forums where community members can feel listened to.

But if the responsible petrochemical facility attempts to secure the goodwill of neighboring communities by making sure neighbors benefit from the presence of the plant and feel that their concerns are heard, it also constructs issues of community health and environmental quality—and engineers' and scientists' authority over them—in ways that undermine residents' claims about how petrochemical emissions affect their health. The inclusion of "social" issues in the mandate of the responsible petrochemical facility allows plant managers to see community criticisms of environmental performance and health impacts not as technical challenges but as failures of communication—failures that can be acknowledged and addressed without undermining managers' claims to technical competence.

Further, bolstered by their status as agents of responsible companies, petrochemical industry engineers' and scientists' constructions of themselves as reasonable, enterprising individuals intersects with residents' own understandings of responsible personhood to help industry's contestable knowledge claims go uncontested. As helpful contributors to Community Advisory Panels, petrochemical industry experts frame technical issues in technocratic ways—and leave little room for community members to advance their understandings of the issues without disrupting the forum's egalitarian structure. Industry engineers and scientists gloss over the considerable uncertainties and extensive unknowns that characterize scientific understandings of the effects of chemical emissions on human health by asserting their own status as responsible choosers who would not work in a place where exposures could be harmful; for residents stuck next to a petrochemical plant, experts' assurances, and the limited data that back them up, can help to preserve not only residents' sense of themselves as enterprising individuals but also the image of their community as a good place to live.

Thus, while the lesson that Karla Raettig learned from her involvement in New Sarpy—that companies will go to great lengths to manipulate communities and defeat community opposition—should be taken very seriously, the analysis here offers an additional lesson: such raw exercises of power are a last resort for most companies, at least in the United States. Rather than cutting down or rolling over community opposition, corporations with resources will work to accommodate a wide range of community concerns and complaints. In doing so, though, they transform their substance: allegations of unjust hiring practices become vocational education programs, for example, and critiques of frequent flaring are answered by better and better explanations of flaring's function. The transformations turn criticism away from areas in which petrochemical companies claim expertise—namely, plant performance, operational safety, and environmental impacts—and shepherd it into areas where companies are willing to grant outsiders a voice.

Accommodating enough to satisfy most community complaints but strategically refusing to address the central issue of community health, petrochemical facilities' response to the criticisms of the environmental justice movement can be thought of as what Marxist theorist Antonio Gramsci called a "passive revolution."[2] Using the concept to analyze Belgian housing policy, geographer Maarten Loopmans and colleagues explain it as follows:

> Passive revolution can be understood as an attempt to re-establish the coherence of the hegemonic project without radical alteration. Policy discourses and practices are reformulated and changed, with three important effects: the incorporation of counter-hegemonic forces' leaders and movements; partial responses to counter-hegemonic claims; and finally, the partial discursive concealment of movements' claims. All three effects globally result in the undermining of the constantly fragile unity of counter-hegemonic forces while re-enforcing hegemonic coherence.[3]

The "hegemonic project" might be thought of as maintaining the dominance of a common worldview in a diverse society—in this case, the idea that refining and petrochemical manufacturing can be undertaken in close proximity to residential communities and, with proper management from technical experts, not pose an undue threat to the health or safety of residents. "Counter-hegemonic forces" comprise those who would question the hegemonic idea—like, in this case, the environmental justice movement. Composed of community groups like CCNS, regional nonprofits like LABB, national and international networks of activist groups, and scholar-activists from both the sciences and social sciences,[4] the movement argues not only that the placement of hazardous facilities next door to poor communities of color is inherently unfair but also that the dangers of the petrochemical industry cannot be justly managed by scientists and engineers without residents' involvement in identifying, framing, and producing knowledge about environmental health issues.

Applied to this case, Loopmans et al.'s characterization of passive revolutions helps to make visible the consequences of petrochemical companies' partial acknowledgment and uptake of environmental justice criticisms—not just for battleground communities like New Sarpy but also for allied groups fighting the idea that environmental hazards can be made acceptable through technocratic approaches to managing risk. The critique of science, scientific expertise, and expert dominance of environmental decision making that is integral to the environmental justice movement's "counter-hegemonic" claims is what goes unanswered in the "partial responses" and what gets obscured in the "partial discursive concealment" that Loopmans et al. describe.

In the face of this partial accommodation, community groups like CCNS may be willing to sideline their assertions that petrochemical pollution is harming their health in order to benefit from much-needed social programs and enjoy cordial relations with plant managers—especially when they see genuine improvements in plant performance, as New Sarpy residents did with Valero's purchase of Orion. This setting-aside represents not a repudiation of their critique of expert knowledge—indeed, several of the residents I spoke to in 2007 continued to refer to illnesses and cancer deaths in the community as evidence that there were continuing dangers from petrochemical emissions—but a strategic decision to fight the fights that are winnable. Community members' latent criticisms of expertise and claims to local knowledge are, further, likely to resurface when and if relations between residents and the refinery reach another crisis point, as they did in New Sarpy not only in 2001 but also in 1996, around the proposed coke conveyor, and 1982, with the construction of the storage tanks.

Should the time come when they want to act against the refinery, New Sarpy residents would almost certainly make common cause once again with environmental justice nonprofits. But in the short term, the nature of industry's partial accommodation tends to undermine alliances between communities like New Sarpy and groups like LABB, which are deeply invested not only in helping communities meet their goals but also in demonstrating the inadequacy of experts' treatment of environmental health issues in fenceline communities. With petrochemical plant managers willing to address criticisms of anything but their technical authority, environmental justice nonprofits may also end up setting aside their critiques of science—as LABB did in Diamond in order to help Concerned Citizens of Norco win relocation.

On the other hand, they may be simply unable to maintain alliances with community groups willing to be satisfied by socially responsible petrochemical experts—as happened in New Sarpy. In either case, environmental justice groups are left to develop their critiques of expertise separate from the immediacy of community campaigns and to try to advance them without the benefit of community members' powerful calls-to-account.

The consequences of industry experts' passive revolution are similarly great for scholars who, like me, strive to use their analytical tools and their empirical insights to make a case for the practical need and philosophical imperative to democratize environmental health science and policy decisions based on it. If environmental justice activists at the community and nonprofit levels accept the bracketing out of science entailed in the petrochemical industry's partial accommodations and choose to fight industrial facilities on grounds of distributive justice, economic justice, and racial justice alone, some kinds of researchers will still be able to be advocates, but the anthropologists, sociologists, political scientists, and philosophers who study the way science is made and used will have far less to contribute to activists' efforts to articulate their critiques and see them translated into policy. Further, without concrete examples of critical, community-based knowledge production, our case for the validity and necessity of diverse ways of knowing becomes an abstract exercise, far less likely to be persuasive to policy makers.

The question, then, is how to resist. How can a critique of expertise continue to be advanced in ways that support fenceline communities' quest for environmental justice in the face of industry experts' neoliberal passive revolution? How can avenues be left open for communities to pursue or return to their situated claims about the health effects of petrochemical emissions while still benefiting from neoliberal social programs? What are the strategies that groups like LABB might employ to maintain critical, community-based knowledge production as an integral part of their work with communities? And how can scholars not only support the efforts of communities and environmental justice organizations but help them travel in ways that can ultimately influence the way environmental decisions are made?

Evolving Expertise in the Environmental Justice Movement

As part of my 2007 visit to New Orleans, I had arranged with Anne Rolfes to spend several days at LABB's offices, helping the organization to craft a basic "Memorandum of Understanding" that would clarify mutual expectations in their work with community groups, and to do a series of interviews with LABB staff and volunteers about their use of sophisticated, real-time air monitors in Chalmette. Walking into LABB's offices, housed in one side of a residential duplex on Canal Street, I got my first glimpse of one of the new monitors that I had come to ask about. A gray, rectangular box nearly two feet long, with several more feet of flexible silver ducting looped back over it, a UV Hound perched on its open, steamer trunk–like case. LABB had rented the instrument, which could measure nearly instantaneously ambient air levels of many of the chemicals tested for in bucket samples, to get a picture of air quality in several of the region's fenceline communities, including New Sarpy, Norco, and Chalmette. But it was not the Hound that most captivated me as I reacquainted myself with LABB. On a high stool in the corner of the front room sat a bucket unlike any I had seen before. Indeed, I was not sure whether I should even call it a "bucket": instead of an opaque, five-gallon paint bucket, the device had an outer shell that was a similarly sized clear food storage container, which offered a clear view of the community-friendly air sampler's characteristic Tedlar bag within.

The two instruments—the transformed bucket and the portable, real-time monitor that had not even existed when we held our monitoring fair in 2002—were emblematic of the growing technical sophistication not just of LABB but of environmental justice organizations across the country. Working with the Hound's designer, middle-aged white engineer Don Gamiles, LABB and a number of other groups were using the portable monitor and its expensive, full-scale cousin, the Sentry, to generate large volumes of data about chemical levels in fenceline communities around the country; the Hound that I saw that day was in fact rented from the Texas-based Community Inpower and Development Association. And the clear food storage container had become a new standard for bucket construction as a result of nonprofit Global Community Monitor's (GCM's) work helping communities all over the world to establish their own bucket brigades: charged with finding local suppliers for bucket components, one Indian community used a container with built-in windows, solving a range of problems associated with the bucket's original design, which included a Plexiglas-covered hole in the lid to allow users to see whether the sampling bag was filling properly. Their innovation inspired GCM founder Denny Larson to look for a transparent alternative to the paint bucket, and organizations in GCM's orbit incorporated the modification when they built new samplers.

At the same time, the juxtaposition of the bucket's new, transparent design, and the Hound, the epitome of a black box, also symbolized significant ambivalence on the part of environmental justice activists in their approach to expert claims about industry's effects on environmental quality and human health. On the one hand, devices like the Hound and the Sentry increased their ability to confront experts on their own terms—to document directly, for example, violations of Clean Air Act standards. On the other hand, long-time organizers like Rolfes were keenly aware that direct action, community mobilization, and media attention were their most powerful weapons against petrochemical facilities, and the easier-to-use bucket better supported those strategies. As Rolfes put it, "You consider you can keep the [Hound] for a week for five hundred dollars or process one [bucket] sample for five hundred dollars, you're getting a ton more data. But a little old lady and her bucket is priceless."

But as activists develop their strategies for challenging experts' claims about, and framings of, the environmental and health issues facing fenceline communities, they not only deal with the problem of whether and to what extent to engage in scientific debates, in which they are at a disadvantage from the start; they confront also the neoliberal ideologies and practices that made engineers and scientists' authority over the issues so robust in St. Charles Parish. Over the years since I last visited LABB, the environmental justice movement's tradition of air monitoring has given rise to three notable initiatives with the potential to heighten challenges to expert knowledge claims. In their various approaches to the problem of how to constitute relations between industry engineers and scientists and the people who live on fencelines of industrial facilities, none stands outside neoliberal ideology; rather, they engage it to varying degrees, more or less critically. What remains to be seen is which, if any, of their strategies for appropriation and critique can broaden the space for democratic participation in science and policy. Will they be able to offer communities like Norco and New Sarpy a resource for continuing their critiques of industry expertise—or taking them up again in future campaigns? More fundamentally, will they be able to shift some measure of authority over environmental health issues away from those with technical degrees and toward those whose claims to knowledge are based on long experience living and breathing in fenceline communities?

Public-Private Partnerships for Air Monitoring

From the earliest days of community-initiated air monitoring, environmental justice activists combined bucket sampling with calls for ongoing, real-time monitoring at refinery fencelines; buckets, they argued, offered some data where there had been none, but a continuous stream of data that could tell residents what they were breathing at any moment was the "gold standard" to which activists thought industry should be held accountable. Fenceline monitoring of this sort remained rare, however, until the mid-2000s, when activists and community groups began working with Don Gamiles, who saw fenceline communities not only as potential users for the real-time air monitors that his company produced but also as the people who could persuade industrial facilities to adopt—and pay for—the sophisticated technologies. While he initially partnered with environmental justice nonprofits—for example sending the first prototype of the Hound to LABB to try out—he has increasingly come to work directly with grassroots community groups, some of which turn to him as an alternative to the confrontational (in their view) tactics of groups like LABB and GCM. In setting up industry-funded air monitoring programs for communities, Gamiles thinks of himself as an honest broker, capable of speaking to both fenceline communities and experts inside the fenceline and dedicated not to proving a particular point but to getting good information.

When I spoke to Gamiles in 2010, he was working on several systems at refineries in the San Francisco Bay Area; one facility had even asked him to write up what they were doing in such a way that it might get taken up as a standard of practice by other sites of the multinational company. But the project that Gamiles was most excited about was the fenceline monitoring system at the Valero refinery in Benicia, California. While he bragged about its technical capabilities—he was installing the most up-to-date monitors available for a wide range of pollutants—the part that was most remarkable, he told me, was that "everyone's getting along." Not only that, but in Benicia they were trying a new model, whereby Valero would pay for the monitors and the costs of operating them for two years, then turn the system over to the community. The idea was that in those two years, the town would have figured out how to integrate the monitoring into other public projects—high school science education, for example—and would have developed new sources of funding to replace Valero's contribution.

As a partnership between a petrochemical company and a local government, designed to promote community entrepreneurship in the long term, the air monitoring project in Benicia represents knowledge production in a neoliberal mode. In general, neoliberal policies, including the rollback of public funding for research, are thought likely to harm science, especially in public-interest areas like environmental health. But scholars have also suggested that neoliberal trends may be generative as well—that the new kinds of partnerships that they promote may open new or previously underdeveloped areas of research.[5] Could, then, the partnership for air monitoring in Benicia be generative of new insights into the effects of petrochemical emissions on community health? More importantly—at least from the standpoint of efforts to democratize science and environmental decision making—could

the community's ownership of so much air quality data enable them to challenge industry experts' framings of environmental and health issues?

In theory, it seems possible. An enterprising community working with sympathetic researchers could, for example, correlate the real-time data with environmental health monitoring, to investigate connections between epi-sodic releases and health impacts—and possibly challenge industry experts' implicit claim that the chemicals they release in flares and other nonroutine circumstances somehow do not count. But Gamiles's reports from the early stages of implementation suggested that dynamics similar to those described in this book are being reproduced around the pioneering project: the group was stuck on how to present the data to the community; Valero wanted to make sure that the data did not cause undue alarm. Their answer was to turn to preestablished standards that would put measurements in context for community members—accepting and further codifying expert understand-ings of environmental health rather than seeing the production of improved, participatory knowledge as a possibility and goal.[6] Community members' apparent commitment to being reasonable and largely avoiding the kind of conflict through which communities can gain power in negotiations with industry also tends to indicate that the refinery's willingness to install air monitors, but not to call into question accepted understandings of what their data mean for community health, is simply an extension of the passive revo-lution described here.

Yet it is possible to imagine circumstances that would prompt the com-munity to adopt a more confrontational stance toward the refinery—a sig-nificant accident, for example, or an expansion that threatened residents' sense of their community as a nice place to live. Under these circumstances, the mass of data generated by the monitoring program could be a resource for residents wishing to challenge experts' claims about health and environ-mental quality; with the help of sympathetic scientists or statisticians, for example, it might be interpreted to show a pattern of low-level but consistent releases or to highlight a dangerous multiplicity of chemical exposures. Such a strategy would call into question expert authority in a manner similar to, but potentially more effective than, bucket monitoring—and entail the cor-responding risks to community image.

A Return to Senses

While some communities have been pursuing ever more sophisticated air monitoring through partnerships with industrial facilities, others have been turning away from monitoring. In the summer of 2007, when I was conducting interviews about environmental justice activists' use of new real-time air monitors, I got to know Ohio Citizen Action, a regional environmental justice nonprofit that had helped communities use both buckets and the Hound as part of "Good Neighbor Campaigns" around Ohio. But by the beginning of 2008, when I approached the group in search of community-based projects for engineering students in an environmental justice class,[7] they were contemplating a new direction—what organizing director Paul Ryder dubbed "three senses monitoring." The Hound *and* the buckets were too expensive, argued the white, middle-aged, long-time activist, and did not let enough people participate. On the other hand, all the eyes, ears, and noses around a facility (he was working with ten neighborhoods surrounding a steel mill in Cleveland) constituted tens of thousands of free monitors, and capturing "data" from them could be a way to increase the number of neighbors involved in the campaign. Ryder proposed—and my sophomore-level undergraduates valiantly tried to implement—a system that let citizens report a noise, sight, or smell by phone, text-message, or web-based form, then compiled the reports on an interactive map.

While the Three Senses (which became Five Senses and ultimately Neighborhood Senses) project never got off the ground in Cleveland, LABB used a similar model in its response to BP's massive oil spill in the Gulf of Mexico in the spring of 2010. In partnership with students from Tulane University, the organization launched the Oil Spill Crisis Map, a website that uses the Ushahidi Platform, open source software originally created to document violence surrounding the 2008 elections in Kenya,[8] to do exactly what Ryder envisioned: allow people to use their mobile phones to report on the environmental consequences of the spill in Gulf Coast communities. By the end of 2011, the map catalogued not only reports related to the BP spill but also reports of spills, flaring, and other releases from petrochemical facilities across southeastern Louisiana.

Ryder's Three Senses proposal and LABB's Oil Spill Crisis Map both fly in the face of typical, expert-driven ways of understanding the impacts of industrial operations. Rather than locating knowledge production in laboratories, or even with high-tech instruments deployed in the field, they see the compilation of discrete observations by diverse individuals as a legitimate, even powerful, way of knowing. Nor are they alone: such "crowdsourcing" is increasingly popular as a means of generating information and even collecting scientific data.[9] But, unlike air monitoring projects, environmental justice organizations' crowdsourcing initiatives do not aim primarily to intervene in scientific discourses by producing information that technical experts necessarily find valuable or even recognizable. Instead, their first concern is to organize: to expand public concern around an issue by giving people a way to participate, and to create graphical representations of a problem around which concerned individuals can mobilize.

Projects like the Oil Spill Crisis Map could thus represent a strategic disengagement from debates with experts on issues of environmental quality and human health. Broadening participation and creating new touchstones for activism in aggrieved communities, they would strengthen direct action, increase public pressure on companies, and make them more likely to negotiate with community groups—without the companies necessarily having to concede questions about the harms done by industrial emissions. They could thus underpin victories like that in Norco, but not change underlying inequities in whose knowledge counts in debates over environmental health.

The potential for more fundamental change lies with the possibility that crowdsourcing of this nature could become accepted as a legitimate, alternative epistemology; that is, sites like the Oil Spill Crisis Map could come to be recognized by, say, environmental regulators (if not petrochemical industry engineers) as an important contribution to knowledge of industry's effects. Were that to happen, even communities who saw themselves as collaborators or partners with industry might participate, using tools like the Oil Spill Crisis Map to maintain a record of local facilities' impacts. But activists and regulators would face the task of making sure the challenges to expertise inherent in such efforts were made explicit by, among other measures, structuring citizen reports in such a way that they could be compiled into larger understandings that did not simply replicate experts' preexisting framings of the issues.

The Refinery Efficiency Initiative

LABB's use of the Hound during my visit in 2007 marked the beginnings of a transition in the way the organization worked with communities. While the organization was still working with St. Bernard Citizens for Environmental Quality on their campaign to get key environmental improvements at the nearby Exxon and Murphy oil refineries, and with the Lower Ninth Ward and Holy Cross neighborhoods to pursue a sustainable model for rebuilding after Hurricane Katrina, its plans for the Hound were broader than these two communities. By sampling in neighborhoods all over the region—including not only Chalmette but places where they had formerly been active (Norco, New Sarpy), as well as fenceline communities where campaigns had not yet coalesced (Garyville, Convent, Alsen)—LABB hoped to create a picture of problems and trends across fenceline communities, partly so that communities might find ways to make common cause.

In the years that have followed, LABB has realized those goals not with air monitoring but through its Refinery Efficiency Initiative. The project, for which the organization received funding from the EPA, collects data from refinery "upset reports" filed with the Louisiana Department of Environmental Quality and compiles them into a searchable, online database. In 2009 and again in 2010, LABB also issued reports that synthesized the information and made a series of claims about Louisiana refinery's record of safety and accident prevention. In *Common Ground II*, they argue that refineries underreport their accidents and do not investigate them thoroughly; they also point to root causes that underlay a large proportion of accidents, including poor storm preparedness and deferred maintenance, that could be addressed to prevent accidents.[10] The Refinery Efficiency Initiative further suggests that refineries would be better able to reduce accidents if they were to collaborate—with one another, with regulatory agencies, with workers, and with communities.[11] In a 2009 letter, LABB and a list of community groups invited refinery managers and their "technical people" to a roundtable, cohosted by the EPA, on "How Stakeholders Can Work Together to Reduce Refinery Accidents":[12] the 2010 *Common Ground II* report notes that twelve of Louisiana's seventeen refineries "have refused repeated invitations to collaborate in good faith" and concludes that "the refining industry is not capitalizing on this opportunity to collaborate to solve the accident problem."

To a greater extent than air monitoring programs, especially real-time systems that characterize everyday air quality, the Refinery Efficiency Initiative appears to challenge refinery engineers and scientists on their own, neoliberal terms. Responsible petrochemical experts like Randy Armstrong and Ellen Williams would be the first to tell you that a large part of their commitments to the environment and the community is preventing upsets, flaring, and other unplanned releases; the LABB's reports make the case that they are not living up to these commitments. Indeed, they offer to *help* the refineries make good on their commitments, proposing in their invitation to the roundtable discussion to bring managers of facilities with a stellar record together with those from plants that need to improve,[13] and suggesting that workers can help plant management address "the most significant causes of refinery accidents: storms and old equipment."[14] This assistance, further, is couched as "collaboration," an idea that arguably plays on the spirit of "communication" and "dialogue" embraced by responsible refiners—as well as by communities wishing to be reasonable, who might find participating in such discussions a way to pursue criticisms of industry performance without incurring the costs of full-on collective action.

Thus the Refinery Efficiency Initiative actually uses ideas integral to the neoliberal reconstruction of refinery expertise, namely, responsibility and communication, to reassert the need for citizen participation in examining and improving refinery operations, an area guarded by the petrochemical industry as an expert domain. For that reason, it marks an important, if subtle, evolution of environmental justice activists' strategies. What will be interesting to see is whether the initiative does indeed provide an avenue for communities like New Sarpy to extend their critiques; whether it is successful in getting refineries to allow citizen participation to influence their practices (the *Common Ground II* report, suggesting that most refineries have refused to participate in the collaborations LABB proposes, does not seem a hopeful sign); and whether a challenge that needles refinery experts' self-constructions as responsible and committed will push them to further revise or refine the grounds on which they assert their technical authority.

Justice's Allies

These recent strategies adopted by environmental justice activists all retain a critical stance toward expertise, thereby resisting the partial accommodation entailed in petrochemical industry responses to environmental justice critiques. They suggest, variously, that industry engineers and scientists' reassurances that air emissions are harmless require measurements—not just models—to back them up; that the impacts of petrochemical pollution are more extensive than experts' narrowly defined air monitoring programs can account for; and that responsible engineers' best operating practices still result in significant numbers of dangerous accidents. The strategies differ in the degree and pointedness of their criticism, just as they differ in the extent and manner in which they adopt the neoliberal ideas central to petrochemical industry experts' renewed authority. Indeed, the two are arguably related: in monitoring programs conceived as partnerships among industrial facilities, community groups, and local government, experts' technical claims are subject only to verification, while crowdsourced maps of the impacts of pollution are able to call into question not only experts' knowledge claims but their very ways of knowing by adopting an explicitly oppositional stance in which partnership and dialogue would seem to have little place. Situated somewhere between those two extremes, the Refinery Efficiency Initiative, with its calls for *increased* partnership and responsibility, suggests the possibility of a strategic deployment of experts' own neoliberal constructs.

Activists' diverse interventions prove less able to resist the wedge, driven by the reconstruction of expert authority, between grassroots groups advocating for improved conditions in their own communities and environmental justice organizations questioning the sustainability of petrochemical production in close proximity to fenceline communities in general. All offer tendrils of possibility for continued community critique of petrochemical industry claims about health and environmental quality. However, in their current forms, none yet incorporate the kind of unified, multilevel assault on expert claims enabled by, for example, buckets in Norco, while simultaneously having the potential to subvert new, neoliberal constructions of expertise. With their emphasis on community-industry partnership, ambient air monitoring programs make an end run around activist groups that would offer a broader critique of petrochemicals, and would require will and significant technical savvy on the part of community members to reinterpret data being gathered. The Refinery Efficiency Initiative, which simultaneously adopts and challenges neoliberal ideas, is not yet linked to organizing in any particular community—although it is conceived in such a way as to be a potential resource and possible showcase for grassroots groups angry about poor safety records at nearby facilities. Only crowdsourced map-making endeavors engage community members directly—but depend on residents' willingness to adopt a confrontational stance, with all of the costs that entails for neoliberal subjects. For any of these initiatives to become a force at the grassroots level, then, it seems that they will require some revision in collaboration with community groups—a real, but unrealized, possibility.

Environmental justice activists have thus not ceased to be combatants in the struggle to characterize the environmental and health harms done by the petrochemical industry, despite industry experts' attempts to win over communities with commitments to social responsibility and appeals to residents' own enterprise—despite, even, industrial facility managers' success in transforming communities like New Sarpy from battlegrounds into friendly territory. To continue to press their critiques of industry knowledge claims, however, activists have (in various ways and to varying degrees) transformed their stances toward science, reconfigured relations with community groups, and even adopted elements of the industry's favored neoliberal logics.

But what of scholars similarly committed to environmental justice? Can they also remain in the fight? That is, in advocating for the democratization of science, technology, and environmental decision making as an integral part of environmental justice, science and technology studies (STS) researchers have depended in no small part on communities like New Sarpy to show what is lost when local knowledge is not incorporated into public policy, and to illustrate the inequities produced when the authority of experts is taken for granted on matters of community health and environmental quality.

When those communities become partners of industry and cease to asso-ciate with the nonprofit organizations through which academics often gain access to them, scholars become detached from some of their strongest allies in efforts to democratize science. What reconfigurations do industry's pas-sive revolution, and activists' responses to it, demand of the academic advo-cates of environmental justice if we too are to continue in our struggles?

Here we might learn from the form that activists' efforts to democratize science have been taking. Each of the strategies detailed above broadens citizen participation in knowledge production: crowdsourcing is the most populist of them, inviting anyone with a cell phone or internet connection to contribute to shared knowledge about the effects of industrial pollution, but collaborative monitoring programs and the Refinery Efficiency Initiative also involve community members in discussions that bear directly on what and how we know about petrochemical effects, including decisions about what technologies to use to measure air toxics levels, how to make sense of monitoring data, and how to interpret and address patterns of accidents at oil refineries.

As efforts at democratization, though, these strategies are unusual in the degree to which they address private sector decisions and decision makers rather than the deliberations of government agencies. That is, the attention of scholars advocating for the democratization of science and policy has for the most part been focused on state-oriented processes: they have argued for changes in the way governments structure public consultation on con-troversial issues of science policy;[15]conducted and assessed experiments in "consensus conferences" in terms of their influence on policy decisions and on the polity;[16] and been participants in community-based research meant either to enhance regulatory science or serve as an alternative to it.[17] Only rarely have these researchers regarded private sector decision making or industry-sponsored deliberations on scientific practice as potential loci for the democratization of science and policy[18]—even as STS scholars have begun to theorize the effects that neoliberalism's privileging of the private sector are having on the practices and institutions of science.[19]

What activists' recent strategies suggest is that the interventions, and not just the analyses, of academics who wish to advocate for environmental justice need to engage new, neoliberal structures of knowledge production, participation, and decision making. In particular, as various functions of government are shifted from the state to the free market, our various efforts at democratization—from participatory research projects to experiments in deliberation—should consider powerful private-sector actors at least as important an audience and target of influence as government agencies and elected officials have historically been. Indeed, to the extent that regulators rely on industry to supply information about its own operations and effects, companies should even be regarded as possible participants in community-based research. Committing to engage the private sector as part of our struggles to democratize science will no doubt require strategically adopting some of the driving ideas of neoliberalism, as environmental justice activists have—even, potentially, ideas that our research gives us reason to be critical of. But if, for example, community-industry dialogue is the primary, state-sanctioned medium available to residents of fenceline communities who wish to participate in decisions affecting environmental quality in their neighborhoods, then it behooves us to find ways to deploy (and appropriate and redefine) the language of "dialogue" in ways that open apparently "technical" issues to public discussion and debate. If the health of communities is to be ensured through the responsible action of multinational companies rather than the protective intervention of the state, then we would do well to illustrate the ethical limitations of "responsibility" as defined by corporations, and to work with activists to articulate those limitations in terms that can help fuel community campaigns.

Many academics are surely already party to advocacy that resists neoliberal transformations, in one way or another; these projects are no doubt both highly worthwhile and fraught with the same tensions and contradictions that attend all forms of public scholarship. Yet, especially in the context of efforts to democratize science and the neoliberal reconstructions of expertise that they have in part engendered, it is important to make explicit the contours of what we are up against and what is required. Having done so, we may be more purposeful in our ongoing struggles to be thoughtful and effective allies to environmental justice activists and the communities who are the victims of environmental injustice.

* * *

With a show of hands in a crowded, windowless, cinder-block room on December 18, 2002, New Sarpy residents ended their community's tenure as a battleground. They went back to being twenty-first-century, neoliberal subjects living with a refinery and forging relationships with the people who managed it, people on whom their health and safety depended. In so doing, they brought into sharp relief what it had cost them to play the role of David battling a corporate Goliath, what damage their eager allies had done to the landscape. They made it possible to see the work that the engineers and scientists on the other side of the fence had done to secure their goodwill—work that not only played on residents' senses of self and community but that also involved experts' own identities. They called our attention, finally, to the ongoing ambivalence that attends living with a responsible, yet inherently hazardous, neighbor. These many years later, I offer their story as a contribution to our collective understanding of the way neoliberal practices and ideologies are remaking science, sidelining issues of social justice, and altering the possibilities for democracy. I hope that it will also serve as a contribution to thinking about how we, as scholars, can promote the democratization of science even when the battlegrounds are quiet, and stand ready to support grassroots groups and their activist allies when ambivalence flares into opposition once again.

CHAPTER 1

1. This outcome is consistent with findings that suggest that community campaigns are more likely to be successful when residents are taking on a proposed facility rather than an existing hazard, and when their struggles attract the attention of national environmental organizations and press coverage. See Roberts and Toffolon-Weiss 2001, Toffolon-Weiss and Roberts 2005.

2. A pseudonym. In this book, both pseudonyms and real names are used, according to each individual's preference. The first occurrence of each pseudonym is designated by an asterisk.

3. Throughout the book, as I introduce new people into New Sarpy's story, I will indicate their race (using the local parlance of "black" and "white"), approximate age, and gender (where it is not evident from their names) as a way of keeping these elements of difference present even though they are not the focus of my analysis. Explicit discussion of race in New Sarpy's evolving relationship with the neighboring refinery can be found in chapter 3, where I show how race, racism, and narratives of the changes in race relations over time played into residents' understandings of community quality and community improvement.

4. Carter told his story to me in an interview in May 2003, several months after the settlement.

5. As the contrast between Landry's assessment of the settlement and that of Winston and Mitchell indicates, "the community" was not always of one mind when it came to Orion. Indeed, intra-community division characterizes many environmental justice campaigns (see, e.g., Roberts and Toffolon-Weiss 2001), and chapter 3 will show how the outcome of CCNS's campaign hinged on a fracture between groups of residents committed to conflicting models of community improvement.

6. Widely used by social scientists, the term "neoliberalism" refers to a suite of political and economic policies that elevate the free market—through, for example, the deregulation of industries, removal of barriers to international trade, and privatization of social services like education—while dismantling the welfare state (see, e.g., Harvey 2005, Jessop 2002). Neoliberalism's free-market rationality also structures notions of citizenship by "extending and disseminating market values to all institutions and social action" (W. Brown 2005, 40; see also Peck and Tickell 2002, Rose 1999, Shamir 2008). While neoliberalism is a sweeping project, it manifests unevenly in particular locales (Brenner and Theodore 2002)—making studies like this one, which examine the situated consequences of both neoliberal policies and neoliberal rationality, essential to understanding neoliberalism's effects.

7. This account of residents' activities during the SEED toxic tour and their subsequent lobbying activities is based on a documentary film that chronicles the events (Dunn 2001).

8. Gray and Szpara 2001, Swerczek 2001 report on the tank fire.
9. The tour's Louisiana stops were reported by Biers 2001, Guarisco 2001, LeBlanc 2001.
10. See Kahn 2001 on the New Source Review issue generally.
11. Lerner 2005 describes Diamond's campaign in detail.
12. See O'Rourke and Macey 2003, Ottinger 2009, Ottinger 2010, Overdevest and Mayer 2008 for a more complete description of the buckets, their history, and their effects.
13. O'Rourke and Macey 2003.
14. Doyle 2002.
15. My account of Richard's participation in the UN Conference on Climate Change is based on the documentary film *Fenceline* (Grünberg 2002); the Shell official with whom she spoke is not identified in the film.
16. See Ottinger 2009, 2010.
17. These include, especially, Fischer 1990, Fischer 2000, Irwin 1995.
18. Guston 1999, for example, discusses a "consensus conference" sponsored in the United States by a consortium of academic institutions and nonprofit organizations.
19. See for example Irwin 2001, Nishizawa and Renn 2006, Rowe and Frewer 2004, Rowe et al. 2004. Irwin's analysis is particularly interesting in that it analyzes UK efforts explicitly informed by the recommendations of social scientists, including Irwin himself.
20. Corburn 2005, Fischer 2000, Hess 2007, Irwin 1995, Martin 2006.
21. American Association for the Advancement of Science 1989, National Science Foundation 1998, Organization for Economic Cooperation and Development 1997, Royal Society 1985.
22. The idea that science is a fundamentally social endeavor characterizes the field of science and technology studies, or STS. Early work in STS closely examined the activities of scientists in their laboratories to demonstrate that scientific knowledge is the outcome of culturally situated practices (see, for example, Knorr Cetina 1981, Latour and Woolgar 1986, Traweek 1988), rather than a set of transcendental revelations about the natural world. A major implication of this work is that science can never be apolitical, nor can it serve as a neutral basis for political decision making. Ongoing calls to democratize technically complex environmental policy decisions thus rest on this foundational tenet of STS.
23. Jasanoff 1990 is one of the earliest works to develop this idea in detail; it has been followed by a proliferation of research showing how science and environmental policy are co-constructed in a manner that stabilizes both (e.g., Miller and Edwards 2001). Bryant 1995, Head 1995, Tesh 2000 offer insight into the uncertainties and contestation involved in the science that surrounds issues of pollution and health in fenceline communities in particular.
24. See for example Funtowicz and Ravetz 1992, Wynne 2003.
25. The case for incorporating local knowledge into science and policy is made in most general terms by Fischer 2000, Irwin 1995; Corburn 2005 proposes a hybridization of local and scientific knowledge that he calls "street science."
26. Irwin and Wynne 1996 deconstruct of the idea of "public understanding of science" using STS concepts and principles.
27. Irwin and Wynne ed. 1996 offers an important early collection of case studies.
28. Wynne 1996.
29. Wynne 1996.

30. E.g., Corburn 2005, Harris and Harper 1997, Johnson and Ranco 2011, Kuehn 1996, Powell and Powell 2011.
31. E.g., Allen 2000, Allen 2003, P. Brown 1993, Brown and Mikkelsen 1997, Corburn 2005.
32. E.g., Brown et al. 2006, Epstein 1995, Epstein 1996, Zavestocki et al. 2002.
33. See Ottinger 2010.
34. A white woman in my midtwenties at the time, I served throughout my year in Louisiana as LABB's "Monitoring Specialist." In that (part-time and unpaid) capacity, I developed tools for interpreting bucket results, gathered information about how industry and government agencies did ambient air monitoring, researched techniques for monitoring that might be suitable for communities, tried to learn how and why residents decided to take, or not take, bucket samples, and organized the Monitoring Fair and Roundtable in New Sarpy—all at the request and under the direction of Rolfes. These activities made me a participant-observer in the aspects of LABB's work that were of most interest, namely, the organization's engagement with expert claims. Through my work with LABB, I also got to know CCNS and CCN members. While residents initially associated me with the environmental justice organization, the association appeared to weaken in their minds when I moved to St. Charles Parish in September 2002 and began asking questions about their community and its history over the course of the fall. In 2003, after the campaign had ended, my involvement in the community—which included interviews, volunteer work at a senior center, weekly attendance at a local church, whose after-school tutoring program and bimonthly hot lunch program I also volunteered for, and attendance at Community Advisory Panels and other public meetings convened by industry—was almost entirely separate from my ongoing, but scaled-back, work at LABB.
35. Epstein 1996.
36. Corburn 2005.
37. Guston 1999, Irwin 2001.
38. The idea that expert authority is fashioned and refashioned by scientists is developed by Gieryn 1999. Gieryn argues that scientists establish their authority over particular domains of knowledge by drawing and maintaining boundaries between science and not-science. In doing so, they make use of the "cultural terrain" of their time, heightening their authority, for example, by positioning science against other domains of activity that are seen as irrational or morally corrupt.
39. See Heynen et al. eds. 2007.
40. Fiorino 2006, Freeman 2006, Karkkainen et al. 2000, Kochtcheeva 2009, Press and Mazmanian 2006.
41. Hoffman 1997 examines the chemical industry's responses to the changing environmental regulatory environment. For further discussion of the Responsible Care program, including critical evaluation of its impacts, see Givel 2007, Gunningham 1995, Howard et al. 1999, King and Lenox 2000, Simmons and Wynne 1993, Tapper 1997.
42. Ward and Dickerson 2001.
43. American Chemistry Council 2001, Lynn and Chess 1994.
44. Harvey 2005.
45. Heynen, et al. 2007 theorize the ways that neoliberalism is changing environmental governance.
46. Miller 2001, Rose 1996b, Rose 1999.

47. Agrawal 2005, Haggerty 2007 analyze the making of environmentally responsible subjects with respect to natural resource management. Neoliberal environmental policies shift responsibilities to individual citizens in other ways, as well—see, for example, Shever 2008, which discusses how Argentinian oil workers have been transformed from employees of the state to owners of small contracting firms.
48. E.g., Holland et al. 2007.
49. E.g., Heynen et al. eds. 2007.
50. E.g., Collins et al. 2008.
51. The idea of cultural terrain comes from Gieryn 1999.
52. Checker 2008, Guldbrandsen and Holland 2001, Holifield 2007, Sawyer 2004, Sze 2007.
53. Holland et al. 2007 offer a particularly useful analysis of these opportunities.
54. In contrast to views that equate expertise with technical knowledge or competence (e.g., Collins and Evans 2002, Collins and Evans 2007), my analysis takes expertise to be a quality based on social relations among people with heterogeneous levels of not only knowledge but also power (Nieusma 2007). Further, it regards an individual's status or authority as an expert as constructed, through processes of boundary work (Gieryn 1999) and the development of scientific personae (Daston and Sibum 2003), among others. Expertise and comparable constructions have been shown to be historically contingent, built on the particular cultural terrain of a time and place (see, for example, Browne 2003, Carson 2003, Daston and Galison 2007, Gieryn 1999, Shapin 1994); my research adds to this body of work by considering how expert authority is fashioned on contemporary cultural terrain. Moreover, by focusing on neoliberal practices and ideologies as defining features of the landscape, it adds to a growing body of work on neoliberalism's effects on science (e.g., Lave et al. 2010, Moore et al. 2011), bringing the reconstruction of expert authority into focus alongside the changes in the funding, organization, and practices of knowledge production theorized by the existing literature.

CHAPTER 2

1. Diona did not specify the man's race.
2. Gray 2000a, 2000b, 2000c.
3. From the "History of Destrehan Plantation," as presented in the video shown to plantation visitors.
4. Gray 2000a.
5. Gray 2000c.
6. When Taylor was a teenager, there was no black high school in St. Charles Parish; her story of having to leave New Sarpy for secondary education is not uncommon among black residents of her generation.
7. The patterns described here were common in both black and white families.
8. Winston and his partner were, as far as I know, the only interracial couple in the St. Charles Terrace neighborhood. Their purchase of land from his partner's family also made Winston the only white person in a black part of the neighborhood.
9. "Close" here is a matter of perspective. Featuring large, two-story homes with grand foyers and multicar garages, and its own golf course, the subdivision in which Armstrong lived was indeed the closest neighborhood in which an affluent, well-educated white person transplanted from elsewhere in the country was likely to consider buying

a home. The plant managers and other high-level officials that worked there thus argued that they lived "in the community," in contrast to other plant workers who lived outside of St. Charles Parish in, for example, New Orleans and its immediate suburbs (Metairie and Kenner) or even upriver in Baton Rouge. The idea that the area the subdivision was in, known as Ormond, was close to the plants had some credibility with New Orleans–based professional activists, as well. On learning that I was living in an apartment complex in Ormond, people like Anne Rolfes either applauded or doubted the wisdom of my choice to live right in the thick of things. Residents of New Sarpy and Norco, however, did not consider Ormond to be close to the plants. Separated from New Sarpy by an undeveloped wooded area, it was generally thought of as out of harm's way. Indeed, when Shell's relocation package allowed Norco activist Margie Richard to move away from her home on the fenceline of the chemical plant, she and her mother bought a house in Ormond, only half a mile or so beyond Armstrong's.

10. See Ottinger 2006 for discussion of conviction and agnosticism on the subject of pollution's health effects.

11. Cole and Foster 2001 offer an overview of the ways in which structural racism produces environmental inequalities. These points are developed further in the extensive academic literature on environmental justice.

12. Allen 2003, Bryant 1995, Head 1995.

13. Allen 2000, Frickel 2008, Hess 2007.

14. Thomas Kuhn makes the point in his foundational book, *The Structure of Scientific Revolutions*; it has since been elaborated by decades of research in science and technology studies (see Hackett et al. 2008). Cohen and Ottinger 2011 situate the insight in the context of science related to environmental health and justice issues.

15. On the idea of local knowledge, see for example Di Chiro 1997, Wynne 1996.

16. E.g., Corburn 2005, Irwin 1995, Fischer 2000.

17. Frickel 2008, Frickel et al. 2010, Hess 2007.

18. Frickel 2008.

19. Rose 1996b; Rose 2001 argues that the enterprising citizen's obligations extend explicitly to protecting and promoting his or her own health.

20. The acknowledgment of obstacles, in fact, puts one into a different political class from enterprising individuals: Rose 1996a argues that the degree to which people can "pass" as responsible choosers determines whether they are "affiliated" with advanced liberal modes of governance or "marginalized" with respect to them. The marginalized, he contends, are governed by a special set of techniques that aim to get people to take responsibility for their lives. This distinction heightens the dangers of New Sarpy residents' strategic stories. By admitting that their responsible choices are structurally constrained, they risk becoming marginal and subject to more coercive forms of governance.

21. For example, the back cover of *Land Sharks: Orion Refining's Predatory Property Purchases*, a report issued by CCNS and LABB in 2002 at the height of New Sarpy's campaign, features a photo of a grand, two-story home, presumably belonging to Orion's pPresident and CEO. The text under the picture reads

> Unlike the people of New Sarpy, the CEO of Orion lives in fresh air, many miles from the pollution and flares of Orion Refining. He lives in a gated community 15 miles away from Orion and the New Sarpy community. If the refinery is as safe as Orion management says it is, why doesn't the CEO live next door?

It is notable that activists here are also telling an enterprising story in order to make the CEO's home into evidence of the refinery's hazards. They presume that the CEO, with all possible choices open to him, has decided to live away from the refinery because he judged neighborhoods nearer the refinery to be less clean and safe than the one he chose.

22. Sensitive to allegations by activists that, if they really believed their plants were so safe, they should live in the neighboring community, the industry scientists and engineers often discussed their residential choices in interviews, in many cases bringing up the subject without prompting from me. A number (about one-third of my small sample) lived in the Ormond subdivision, as Randy Armstrong did; the rest were scattered across the region. While the former group touted their decision to be part of the community, the latter talked about their partners' commutes, children's schools, and assorted amenities offered by different locales in explaining how they came to live where they did.

23. See chapter 4 for an extended discussion of the CAPs.

24. Several parishes, starting with St. Charles Parish on the downstream end, that span the Mississippi River. Together they comprise most of the region known as the "Industrial Corridor" or "Cancer Alley."

25. Minutes from November 20, 1997, suggest that this analogy was meant to convey the magnitude of the problem: "The Lung Cancer rate in Southern Louisiana equaled one plane crashing and killing 220 passengers each month."

26. In echoing the logic of the enterprising individual with a "will to health" (Rose 2001, 6), Chen ignores the complicated, well-documented links between social inequalities—including environmental inequalities—and health (Nguyen and Peschard 2003 offer an overview).

27. Others scholars have also shown how, in a variety of circumstances, marginalized individuals and communities have invested in and mobilized discourses of neoliberalism—and how these strategies have further disadvantaged vulnerable groups (Boyd 2008, Guldbrandsen and Holland 2001, Pérez 2008, Pearson 2009).

CHAPTER 3

1. For more detail on the historic sites in this part of St. Charles Parish, see Sternberg 1996, 117–31.

2. In New Sarpy and nearby communities along the Mississippi River, "front" and "back" are commonly used to indicate toward and away from the river, respectively. These terms are probably left over from a time when large plantation homes fronted on the river, and moving away from the river took you in back of the house. Use of the word "riding" to indicate driving in one's car and the use of definite articles before certain major thoroughfares (e.g. "the River Road," "the Airline") are also common quirks of local parlance that I have chosen to adopt here.

3. It is significant that, in New Sarpy, the line of racial segregation ran perpendicular to the refinery fenceline. Because black families and white families were equally close to Orion, CCNS members included people of both races, and campaign events routinely drew participants from both the black and white parts of the neighborhood. This racial diversity, however, prevented the community from including charges of environmental racism as central to its campaign, as Diamond residents had done to good effect (cf. Checker 2005).

4. Norco is more than twice the size of New Sarpy: in the 2000 Census, Norco had a population of approximately thirty-six hundred compared to seventeen hundred in New Sarpy.

5. The refinery that is now Motiva was a Shell facility until 1999; its units are still intermingled with those of Shell Chemical's east site.

6. Harvey 1989.

7. Harvey 1989, Peck and Tickell 2002.

8. The importance of the public image of, or symbolic meanings associated with, a place is documented in a number of case studies, including Brownlow 2009, Jaffe and Quark 2006, and Rousseau 2009.

9. One of the few studies that deals with this issue explicitly, Brownlow 2009 shows how policing statistics are manipulated to create the image of a city as a safe place.

10. While campaigns for relocation were predicated on the claim that residents were stuck in their neighborhoods because of depressed home prices, properties in New Sarpy and Norco were still bought and sold—though not necessarily for what residents thought they should be worth.

11. Cf. Phillimore and Moffat 2004.

12. St. Charles Parish Economic Development Commission. 2009. *St. Charles Parish: The Best of All Worlds.* http://www.stcharlesgov.net/index.aspx?page=79 (accessed August 24, 2010).

13. Harvey 1989, 8.

14. The gendered language here is deliberate: all five were men, four of them white and one black.

15. Cf. Harvey 1989, Brownlow 2009. These and other authors suggest the importance of quality of life, especially safety, to urban entrepreneurialism; however, they do so mainly in the context of strategies that focus on attracting consumer dollars, largely ignoring the connections between quality-of-life issues and strategies that focus on attracting industrial production.

16. A handful of other chemical companies, including Dow Chemical, also had businesses in Norco. Their sites were relatively small and contained within the footprint of Shell and Motiva. Some of the companies, in fact, had purchased pieces of the larger companies' operations; Resolution, for example, established itself in Norco when Shell sold off its resins business in 2000.

17. "Good Neighbor Initiative Wins Top Award." Good Neighbor Initiative Newsletter, May 2003.

18. Histories of Norco—at least as told by the River Road Historical Society, the organization awarded the grant for Community History Day—usually focused on the way the town grew from a community of white Shell workers in company-provided housing to an independent, off-site community of white families who had worked for Shell for several generations. Diamond residents told a different history, of being displaced from their original community, called Belltown, by the Shell Chemical plant in the late 1950s, excluded from jobs at the plant, and confined to the margins of segregated Norco (see Lerner 2005). Because of these polarized histories, it is hard to imagine that Community History Day (which occurred after the period of my fieldwork) could have been a truly inclusive event.

19. The Norco Christmas Parade, on the other hand, had no comparably controversial message and attracted both white and black residents in droves. Racism—in its

twenty-first-century incarnation—was visible nonetheless. I attended the 2002 Norco Christmas parade with a black friend and her extended family, which included two teenaged girls. The girls were competing with each other for "throws"—the candy, beads, and other trinkets thrown off parade floats, à la Mardi Gras—shouting and jostling as they did, with little regard for the people around them. When a convertible carrying a white beauty queen, also a teenager, passed, the girls rushed up, getting within inches of her as they shoved each other. The white girl recoiled and, instead of throwing a trinket to them, she held one out grudgingly in her open palm. One of the girls with my group snatched it up and ran off with her cousin—not realizing that she had scratched the beauty queen's hand. The white girl glared after them, rubbing her hand—but she probably failed to understand that her reluctance to throw to the girls, as she had to others in the crowd, was partly to blame for her injury.

20. Harvey 1989.
21. Lerner 2005, 11–17.
22. Sternberg 2001, 117–31.
23. Lerner 2005, 135.
24. This could also be interpreted as bringing appraised values up to something more like what homes would be worth if there were not a chemical plant right on top of them. Property in the white parts of Norco was actually surprisingly valuable: in the documentary *Fenceline*, a resident points out a home that had recently been on the market for $277,000. When environmental justice activists demanded to know why one junior GNI staff member, a single mother in her early thirties, did not live in Norco—a question they fired at many industry representatives—she reportedly retorted that she would love to, but she could not afford it.
25. Belonging to a church was an important part of life for many—perhaps most—New Sarpy residents, yet how and where people did so varied quite a bit depending on whether they were black or white. There were four Baptist churches in the town of New Sarpy (though only one of them was in the St. Charles Terrace Subdivision), all of which had entirely black congregations. St. Matthew Baptist Church, where I attended services every Sunday from January until June 2003 (see Ottinger 2006), was by far the largest of these, as well as the oldest. Most of the black families I knew had some connection to the congregation—and several of CCNS's most active members were also active in the church—and most of the people I knew in the congregation had some familial connection to the town, even if they no longer lived there. White people attended church outside of New Sarpy: depending on their denomination, some went to the historic St. Charles Borromeo Catholic church in Destrehan, for which the parish is named; others to the Baptist church in Norco; and a few drove further to other kinds of Protestant churches (I met no one in St. Charles Parish who was not, at least nominally, a Christian), including a giant evangelical church in Metairie.
26. The phenomenon of messy neighbors—and complaints about them—seemed to cut across racial lines. Black and white residents were just as likely to complain to me about the state of their neighbors' properties, and because whites were concentrated in the "front" block and blacks in the "back" of the neighborhood, the messy neighbor was almost always the same race as the complainant. One elderly white man, in fact, told me that he would have been happier if his sister's old home—the house next to his—had been sold to a nice black family that would keep the place up, rather than

the new owner, a young white man who let piles of trash sit in the yard. The contrast he made between responsible, older blacks and indolent young people was echoed by many of the seventy-somethings in New Sarpy, though it had different functions with respect to race, depending on the speaker. White resident Ida Mitchell, for example, posited a generational gap in order to draw a distinction between black residents who had been good neighbors and even friends for many years, and the kind of blacks she had no use for: the young, disrespectful, rap-blaring gang members with their pants hanging off that one sees portrayed in the media. Audrey Taylor, a black resident of a similar age, also complained about the younger generation, but she was concerned about the way that groups of teenagers loitering on New Sarpy's streets, which from my observations usually included both black and white youth, brought down the neighborhood. The one distinctly racist charge routinely made in the context of residents' efforts to keep New Sarpy nice had to do with automobile traffic through the neighborhood. Whites living in the "front" block, closest to the river, would complain about the cars that sped down their not-quite-two-lane street at forty and fifty miles per hour, making a big racket and endangering children who might be playing outside. Because these speeders were said to be coming to and from "the back," the unstated accusation was that black drivers were responsible for this bane in the neighborhood.

27. The report in fact connects Lee to Norco, where much was made of Diamond residents' ancestral connection to the 1811 slave revolt (see Lerner 2005, 13–17; white residents of New Sarpy would not have been described as having a "deep history with the land"). Lee's grandfather relocated to New Sarpy from Belltown, on the far west end of Norco, when Shell bought the land in the 1950s to build its chemical plant. Like Lee, many black residents of New Sarpy had family connections to Norco: some grew up there, others had siblings who went to live near in-laws in Diamond after they were married, many had cousins who lived there. White New Sarpy families did not have similar ties to Norco; in fact, historically, white Norco families were far better off than their New Sarpy counterparts by virtue of their jobs at Shell.

28. Ottinger 2009 describes how bucket results are interpreted to show systemic dangers of living next door to a chemical plant.

29. See note 25 above.

30. This point was made most volubly by Don Winston, who likened the company's heavy-handedness to "the old master-slave relationship." I found the comparison shockingly inappropriate, especially coming from a white man raised in the Northeast. Whatever anyone else may have thought of his comments, though, no one challenged or criticized his rhetoric in my hearing. In fact, that part of Winston's testimony was broadcast as part of the local television station's coverage of the press conference. Yet it did not become a rallying cry or get adopted by others involved with the campaign—suggesting to me that perhaps the comment was one that Winston's fellow CCNS members let stand out of politeness or an interest in maintaining a unified front rather than due to any support for his sentiment.

31. As long as CCNS's campaign was going on, I did not have access to Orion officials or CCNS's detractors in the community: having been introduced to the community through LABB and having established relationships with CCNS's leaders, I was seen as allied with the campaign. My interviews with SCTNA leaders and Orion managers all occurred in 2003, after the controversy had been settled.

32. A seafood boil—or more often, a crawfish boil or a shrimp boil—is a common kind of festive event in south Louisiana, akin to a cookout or a barbeque. Usually held in someone's back yard, a boil features not a grill but a big pot of water, in which seafood and corn on the cob are cooked.

33. The founding group had already held at least one, and perhaps several, meetings at which officers had been selected. The first open meeting was planned for late September but was preempted by a hurricane and subsequently rescheduled for early November.

34. Dividing the neighborhood by streets when the line of racial segregation occurred at the midpoint of *each* street suggests how little SCTNA's all-white board had considered the racial politics of the community in setting up the organization.

35. The effects that relocation would have on the community were hypothetical for New Sarpy residents at this point in CCNS's campaign, but while they were struggling over how to build up their community, many Diamond residents were preparing to move out of theirs. By the following spring, just what a relocation program meant for those who did not care to leave was becoming more obvious. At the end of April 2003, I interviewed George and Harriet Lewis,* a black couple in their sixties who had chosen not to take advantage of the Diamond Options Program. They felt they could not replace their large lot and their rancher house, which they had added onto incrementally over the years, for what the Shell program was offering, and that, in order to really get away from pollution, they would have to move out of the *region* that was their home. Being one of just a handful of families left in Diamond was not so bad, they told me, although they missed their old neighbors quite a bit. They had a lot of green space around them, which they enjoyed; without all the houses around to help absorb the sound, though, the noises from the chemical plant had gotten louder. Harriet Lewis also said their neighborhood was in danger of being neglected, because there were so few people left: "Some of the services we're supposed to get we're not getting because they feel like the people are gone. You know, we have to call up and remind them we have some people still in the neighborhood." The effect of the relocation program was felt beyond Diamond as well: two local business owners told reporter Emily Bazelon that business had dropped precipitously since Diamond residents starting moving out (Bazelon 2003).

36. The establishment of a second community group was clearly in Orion's interests. Though community members would have had no way of knowing, at the time of the seafood boil Orion would already have been in the process of negotiating a settlement with the LDEQ that included fines and beneficial environmental projects, and they would have known that, in order for beneficial environmental project money to flow to the community, the company would have to work with a group that was officially incorporated—something that CCNS never was. Moreover, as Jason Carter would later tell me, Orion officials felt strongly that CCNS was not representing the whole community, and they wanted to find ways to reach those segments of the neighborhood who were interested in working with them. For these reasons, I—like others associated with CCNS's campaign—have always found it hard to believe that Orion did not suggest that Burnham and others at the seafood boil form SCTNA, though I can cite no proof of my suspicions.

37. These included exceeding permitted limits for particulate matter and other air pollutants; emitting excessive sulfur dioxide in accidents and emergency situations,

which are not included in permitted limits; failing to report accidents and emergency conditions adequately; and failing to maintain continuous emissions monitors.

38. Latour 1987 shows how it becomes harder and harder to dissent from scientific claims as allies are amassed behind them; see chapters 1 and 2 in particular.

39. These views are expressed by white Norco residents featured in Lerner 2005.

40. Lerner 2005.

41. Civic Association Sets Standard. St. Charles Herald-Guide, February 19, 2003.

42. Ahmed and Schaeffer 2002.

43. The angry complaints of residents at this meeting not only showed the baldest racial animosity that I saw in my time in Louisiana but also offered a window into the paternalistic role that Shell was expected to play in the community. In 2001, Shell purchased the Gaspard-Mulé tract, the one-block-wide strip of land that separated Diamond from the white sections of Norco, in order to get the riverfront property that was part of the tract. Their purchase caused them trouble at the time: CCN, in the throes of their relocation campaign, asked how Shell could be "buying trees and ignoring our pleas." But by late 2002, having reached a settlement with CCN, Shell was ready to incorporate it into the "greenbelt" that their recently acquired land in Diamond was to become and cleared out the dense underbrush that had grown up on the untended property. At the GNI meeting, residents of Mary Street, the street on the white side of town closest to the tract, alleged that the clearing of the land had made it possible for "those people" to come over (blacks from Diamond, they clearly meant), case their homes, call dirty comments to them, and steal from them. Residents blamed Shell in no uncertain terms for these developments: "You have given criminals easy access to what we have," one woman said to the GNI staff. Further, many said that Shell ought to compensate residents for, according to them, turning Mary Street from one of the best streets in Norco to one of the worst. That Shell was not willing to relocate Mary Street residents, or offer them home improvement loans, as they had done for Diamond residents, caused residents attending the meeting to complain that *they* were victims of injustice in Shell's treatment of the two parts of the community. As one woman put it, "You cater to this group of people and you expose us to robberies and ugly names. You help them out, but you don't help us."

44. See chapter 2.

45. Phillimore and Moffatt 2004, Moffatt et al. 2000 similarly demonstrate the connection between scientific studies and community image with a case study where the converse was true: a scientific study was not taken up because it reinforced a community's image as a polluted place.

46. Latour and Woolgar 1986 (chapter 2) describe how scientific claims become facts by being taken up and used by others; see also Latour 1987, chapter 1.

CHAPTER 4

1. Residents and Industry Should Work Together, St. Charles Herald-Guide, July 20, 2002.

2. Ward and Dickerson 2001, 2.

3. Habermas 1989 [1962].

4. This term is drawn from Mansbridge et al. 2010.

5. Foster 2002, Guldbrandsen and Holland 2001.

6. As part of its environmental justice programs, for example, the EPA promotes the Collaborative Problem Solving (CPS) Model through grants to communities who

propose to use it to address local environmental justice issues. While the model is decidedly community centered, prioritizing community "visioning" and capacity building, it also encourages communities to see local industry as potential stakeholders in consensus-building efforts and to engage in "facilitated dialogue" with companies that may be seen as the source of environmental justice issues in order to avoid "hostility and an extensive legal debate" (United States Environmental Protection Agency 2008, 27).

7. See chapter 5.

8. Fraser 1992, Benhabib 1992.

9. Young 2001 contrasts activist and deliberative approaches to politics; see also Mansbridge et al. 2010.

10. Lerner 2005, 245–60.

11. Lerner does not make reference to Warshall's race, though he mentions that Warshall was sixty at the time.

12. Habermas 1989 [1962].

13. See Young 2001.

14. Ryder 2006.

15. Ryder 2006, 130.

16. Ryder 2006, 135.

17. In fact, Rolfes is quoted at length at several points in Ryder 2006 on the subject of negotiations in Norco.

18. Lerner 2005, 247–48, also describes this event.

19. Fortun 2001, 114; Allen 2003, 93–100.

20. Henry LeBoyd, the black New Sarpy native who was, in his role as Orion's community relations manager, instrumental in setting up Orion's CAP, explained his decisions about the CAP's structure in a May 2003 interview:

    So my thinking is that this place has such a wretched past that to try to incorporate it with a parish-wide CAP with companies that have been stable here for forty and fifty years, would do a disservice to them and probably do nothing for us at all. Okay. So my thing was that we need to go this one alone. And we need to go parish-wide.

    "Why parish-wide?" I asked. He explained, "Because if I stay in the area only where my enemies are, nine chances out of ten I will always be in a fight."

21. Cf. Sawyer 2004.

22. Benhabib 1992, Fraser 1992.

23. Lynn and Chess 1994, Lynn et al. 2000, Ottinger 2008.

24. Minutes from all St. Charles CAP meetings were filed with the St. Charles Parish public library.

25. For discussion of the limits of risk-based framings in the environmental justice context, see Kuehn 1996, Johnson and Ranco 2011.

26. Latour and Woolgar 1986 describe in detail the process through which scientific facts come to stand on their own.

27. See chapter 2.

28. While Ramachandran's comments were not recorded, the quoted turns of phrase are captured in notes taken during and immediately after the panel meeting.

29. Ottinger 2009 discusses these competing frameworks, or evidentiary contexts, for understanding data from air monitoring.

30. Given the volume and complexity of scientific studies on chemical health effects, it is hard to believe that the standards are updated every time new data are released. While no history of revisions to the standards is readily available, the standards have not been revised since at least 2001.

31. Black residents of Norco in attendance at the meeting, including CCN members, also chose not to engage Ramachandran's claim, perhaps because, having already won relocation for the Diamond community, they had little at stake. Had the claim been made during the campaign, however, it seems assured that a CCN member would have challenged it as self-interested and experience-based by pointing out that what Ramachandran presented as certain knowledge was based only on the observations of a few white Shell retirees.

32. I make this proposal in Ottinger 2010 as well.

33. See, e.g., Irwin 2001, Guston 1999.

34. Irwin 2001.

CHAPTER 5

1. The surge in population in St. Charles Parish was a result of Hurricane Katrina, which rendered much of New Orleans and sections of neighboring Jefferson Parish uninhabitable. The storm left communities in St. Charles Parish without power for up to a week but did not cause significant damage otherwise.

2. Merton 1979 [1942].

3. Shapin 1994, Brown and Michael 2002; see also Carson 2003, Secord 2003, Shapin 2003, 2008.

4. On the suppression of data, see, e.g., Infante 2006, McGoey and Jackson 2009, Tong and Olsen 2005; on manufacturing uncertainty, see Michaels 2008, Oreskes and Conway 2010, Shrader-Frechette 2007. Scholars in social studies of science have also documented more subtle mechanisms by which corporate involvement shapes scientific research (see Cooper 2009, Freudenberg 2005, Lave et al. 2010, Moore et al. 2011) and science-based regulation (Abraham 1993).

5. Noble 1977, Reynolds 1991.

6. Lynch and Kline 2000, Swierstra and Jelsma 2006. For an example, see the National Society of Professional Engineers' Code of Ethics (http://www.nspe.org/Ethics/Code-ofEthics/index.html) and their Board of Ethical Review (http://www.nspe.org/Ethics/BoardofEthicalReview/index.html).

7. Shamir 2008, 2010.

8. Shamir 2008.

9. Shamir 2010, Watts 2005.

10. While company-initiated environmental programs made a difference in this case, the overall effectiveness of voluntary initiatives versus environmental regulation is questionable: see Borck and Coglianese 2009, Gunningham 1995, Stretesky and Lynch 2009.

11. http://www.valero.com/default.aspx.

12. http://www.valero.com/OurBusiness/Pages/Excellence.aspx.

13. Shamir 2010, Watts 2005.

14. Fortun 2001, Hoffman 1997.

15. Motiva is a joint venture of Shell and Saudi Aramco; while the company has its own statement of general business principles (see http://www-static.shell.com/static/

motiva/downloads/pdf/code_of_conduct.pdf), it appears to replicate, possibly with minor modifications, those of Shell.

16. Sadler and Lloyd 2009, Oglesby 2004, Shamir 2005, 2010.

17. Zyglidopoulos 2002, Thornton et al. 2009.

18. Shamir 2010 and Zyglidopoulos 2002 both give the example of Shell's decision to decommission the Brent Spar oil storage buoy through deep-water disposal in the North Sea. Although the buoy was under the management of Shell U.K. and regulated by the U.K. government, it was public outcry in continental Europe that led Shell to rethink its decision.

19. The logic of neoliberal governance thus functions in parallel ways with respect to corporations and individuals: the responsible corporation and the enterprising individual (see Rose 1996b) are both manifestations of the self-governing subject at the core of neoliberal governance.

20. Shamir 2008.

21. Shamir 2010 and Sibley 2009 discuss the ways in which even low-level workers are invested with responsibility for the performance of large organizations; Sibley in particular is critical of this move as eliding differential interests and inequalities in power between workers and management.

22. Although I met women in the second tier of facility managers, especially in the role of health, safety, and environment manager, I do not know of a single woman who has served as the top plant manager at the helm of the Norco and New Sarpy facilities discussed here, before, during, or since my fieldwork.

23. Implied in CSR, the idea of a social license to operate is examined explicitly by Gunningham et al. 2004, Howard-Grenville et al. 2008.

24. The locations of the refineries named by this plant manager have been changed to protect his anonymity.

25. Welker 2009 offers a more thorough analysis of this kind of activist-blaming rhetoric as part of the construction of responsible companies.

26. The EPA is organized into ten regional divisions, with Region 6 serving Arkansas, Louisiana, New Mexico, Oklahoma, Texas, and sixty-six tribal nations.

27. See Shapin 2008 for a discussion of the relations among moral authority, credibility, and expertise in contemporary contexts.

28. Ellen Williams's rise to the rank of vice president is just one example of this trend. Wayne Pearce is another: having started as Shell Norco's plant manager in 2000 after twenty years with the company, he was promoted to managing director of Shell & BP South African Petroleum Refineries in 2004; by 2011, he had become vice president of process safety assurance for Shell.

29. Cf. Shapin 1994, 409–17. Here, face-to-face evaluations of personal virtue are not replaced but supplemented by evaluations of institutions. Those evaluations, however, are focused not on the degree of control exercised by the institution but on the imagined virtues of the institution itself.

30. Johnson's testimony to the Parish Council occurred just six weeks after a massive fire in a gasoline storage tank at the refinery that sparked a demonstration by community members outside refinery gates. From the content of his remarks, it appears that the council was, at least in part, calling him to account for the accident.

31. In our interview, Williams explained that Valero had "a whole corporate group of folks" in San Antonio who were experts on particular environmental issues (e.g.,

waste, air compliance) and technologies (like the sulfur recovery unit or the coker), on whom she called as issues arose at the refinery. These experts were almost never visible from the community's point of view.

32. The reason Armstrong gave for feeling badly about the buyout was that many residents would be moving out of Diamond because of their perceptions that environmental risks were unacceptable, into places where crime rates were higher and schools were worse. In his judgment, such a move would not result in better quality of life for former members of the community.

33. Lerner 2005 represents the full range and complexity of residents' grievances, showing them not to be easily reducible to environmental concerns any more than to complaints about Shell's history of racist practices.

34. Social studies of technology and engineering have, in fact, shown that so-called social competencies are integral to engineering work (e.g., Law 1987, Suchman 2000), yet engineering identities continue to privilege technical skills and problem solving—see Faulkner 2007, Trevelyan 2010. Moreover, the socially inept engineer remains a common stereotype in popular and engineering cultures ("How do you know someone is an outgoing engineer?" I was once asked by the president of an engineering college. "He looks at *your* shoes when he's talking to you.").

35. In fact, the belief in the power of communication is so widely shared within the industry that more than one of Orion's top managers came to New Sarpy expecting to do exactly what Ellen Williams described. Having come from other facilities with good relationships with neighbors, they told me, one of the first things they wanted was to establish a "very active dialogue with the community," in Jason Carter's words, of the kind they had experienced in their previous posts; indeed, they scheduled a first community meeting within a few weeks of the new plant manager's arrival and were bewildered when their efforts were met by a hostile crowd and a contingent of reporters.

36. See for example Young 2006.

37. Shever 2010, Welker 2009.

38. See Shrader-Frechette 2002 on the occupational case.

39. Downey 2005, Downey et al. 2006, Ottinger 2011b, Riley and Bloomgarten 2006.

CHAPTER 6

1. Scholars like Melissa Checker (2005) and Steven Gregory (1999) have shown the environmental justice activism of particular communities to grow out of long histories of social justice and civil rights organizing; the view is echoed by accounts of the environmental justice movement in general that locate its origins in the civil rights movement (see, e.g., Cole and Foster 2001). On the other hand, the environmental justice literature also describes cases in which local environmental struggles are residents' first introduction to political action (e.g., Cole and Foster 2001, Pardo 1998). As this book shows, New Sarpy has no deep history of collective action, and there is reason to believe that, as more low-income white and racially mixed communities undertake environmental campaigns, their story will be the more typical one.

2. Hoare and Smith 1971, 119-20.

3. Loopmans et al. 2010, 195–96.

4. Cole and Foster 2001, Schlosberg 1999 discuss the heterogeneous groups that make up the environmental justice movement. On the participation of scientists and social scientists in the movement, see Ottinger and Cohen 2011, Cable et al. 2005.

5. Lave et al. 2010, Moore et al. 2011.
6. See Ottinger and Zurer 2011 for an extended discussion.
7. For more on these projects, see Ottinger 2011b.
8. http://www.ushahidi.com/about-us.
9. Cohn 2008, Dickinson et al. 2010, Ellis and Waterton 2004, Keim 2009.
10. *Common Ground II: Why Cooperation to Reduce Accidents at Louisiana Refineries Is Needed Now*, available at http://www.labucketbrigade.org/downloads/LABBcommon-groundIImap.pdf (accessed April 15, 2012).
11. "Refinery Efficiency Initiative," Louisiana Bucket Brigade, available at http://www.labucketbrigade.org/article.php?list=type&type=169 (accessed April 15, 2011).
12. "Letter to Refinery Plant Managers," Louisiana Bucket Brigade, available at http://www.labucketbrigade.org/article.php?id=490 (accessed April 15, 2012).
13. "Letter to Refinery Plant Managers," Louisiana Bucket Brigade, available at http://www.labucketbrigade.org/article.php?id=490 (accessed April 15, 2012).
14. "Refinery Efficiency Initiative," Louisiana Bucket Brigade, available at http://www.labucketbrigade.org/article.php?list=type&type=169 (accessed April 15, 2011).
15. Irwin 2001.
16. Joss 1998, Guston 1999.
17. E.g., Liévanos et al. 2011.
18. Hess 2007 is a notable exception.
19. Lave et al. 2010, Moore et al. 2011.

REFERENCES

Abraham, John

1993   Scientific Standards and Institutional Interests: Carcinogenic Risk Assessment of Benoxaprofen in the UK and U.S. Social Studies of Science 23(3): 387-444.

Agrawal, Arun

2005   Environmentality: Technologies of Government and the Making of Subjects. Durham, NC: Duke University Press.

Ahmed, Huma, and Eric V. Schaeffer

2002   Smoking Guns: Smoking Flares and Uncounted Emissions from Refineries and Chemical Plants in Port Arthur, Texas. Washington, DC: Environmental Integrity Project.

Allen, Barbara L.

2000   The Popular Geography of Illness in the Industrial Corridor. *In* Transforming New Orleans and Its Environs: Centuries of Change. C. E. Colten, ed. Pp. 178-201. Pittsburgh: University of Pittsburgh Press.

2003   Uneasy Alchemy: Citizens and Experts in Louisiana's Chemical Corridor Disputes. Cambridge, MA: MIT Press.

American Association for the Advancement of Science

1989   Science for All Americans. Washington, DC: American Association for the Advancement of Science.

American Chemistry Council

2001   Guide to Community Advisory Panels. Arlington, VA.

Auyero, Javier, and Débora Alejandra Swistun

2009   Flammable: Environmental Suffering in an Argentine Shantytown. Oxford: Oxford University Press.

Bazelon, Emily

2003   Bad Neighbors. Legal Affairs, May/June: 53-57.

Benhabib, Seyla

1992   Models of Public Space: Hannah Arendt, the Liberal Tradition, and Jurgen Habermas. *In* Habermas and the Public Sphere. C. Calhoun, ed. Pp. 73–98. Cambridge, MA: MIT Press.

Biers, John M.

2001   Environmentalists Air Concerns about Refineries: Protesters Urge Bush to Support Controls. (New Orleans) Times-Picayune, July 28: 1.

Borck, Jonathan C., and Cary Coglianese

2009   Voluntary Environmental Programs: Assessing Their Effectiveness. Annual Review of Environment and Resources 34: 305-24.

Boyd, Michelle R.

2008   Integration and the Collapse of Black Social Capital: Nostalgia and Narrative in the Neoliberal City. *In* New Landscapes of Inequality: Neoliberalism and the Erosion of

Democracy in America. J. L. Collins, M. di Leonardo, and B. Williams, eds. Pp. 91-112. Santa Fe, NM: School for Advanced Research Press.

Brenner, Neil, and Nik Theodore

2002   Cities and the Geographies of "Actually Existing Neoliberalism." Antipode 34(3): 349-79.

Brown, Nik, and Mike Michael

2002   From Authority to Authenticity: The Changing Governance of Biotechnology. Health, Risk, and Society 4(3): 259-72.

Brown, Phil

1993   When the Public Knows Better: Popular Epidemiology Challenges the System. Environment 35(8): 16-21, 32-41.

Brown, Phil, Sabrina McCormick, Brian Mayer, Stephen Zavestocki, Rachel Morello-Frosch, Rebecca Gasior Altman, and Laura Senier

2006   "A Lab of Our Own": Environmental Causation of Breast Cancer and Challenges to the Dominant Epidemiological Paradigm. Science, Technology, and Human Values 31(5): 499-536.

Brown, Phil, and Edwin J. Mikkelsen

1997   No Safe Place: Toxic Waste, Leukemia, and Community Action. Berkeley: University of California Press.

Brown, Wendy

2005   Edgework: Critical Essays on Knowledge and Politics. Princeton, NJ: Princeton University Press.

Browne, Janet

2003   Charles Darwin as a Celebrity. Science in Context 16(1/2): 175-94.

Brownlow, Alec

2009   Keeping Up Appearances: Profiting from Patriarchy in the Nation's "Safest City." Urban Studies 46(8): 1680-1701.

Bryant, Bunyan

1995   Issues and Potential Policies and Solutions for Environmental Justice: An Overview. In Environmental Justice: Issues, Policies, and Solutions. B. Bryant, ed. Pp. 8-34. Washington, DC: Island Press.

Cable, Sherry, Tamara Mix, and Donald Hastings

2005   Mission Impossible? Environmental Justice Activists' Collaborations with Professional Environmentalists and with Academics. In Power, Justice, and the Environment: A Critical Appraisal of the Environmental Justice Movement. D. N. Pellow and R. J. Brulle, eds. Pp. 55-76. Cambridge, MA: MIT Press.

Carson, Cathryn

2003   Objectivity and the Scientist: Heisenberg Rethinks. Science in Context 16(1/2): 243-69.

Checker, Melissa

2005   Polluted Promises: Environmental Racism and the Search for Justice in a Southern Town. New York: New York University Press.

2008   Withered Memories: Naming and Fighting Environmental Racism in Georgia. In New Landscapes of Inequality: Neoliberalism and the Erosion of Democracy in America. J. L. Collins, M. di Leonardo, and B. Williams, eds. Pp. 169-90. Santa Fe, NM: School for Advanced Research Press.

Cohen, Benjamin R., and Gwen Ottinger

2011   Introduction: Environmental Justice and the Transformation of Science and Engineering. In Technoscience and Environmental Justice: Expert Cultures in a Grassroots Movement. G. Ottinger and B. R. Cohen, eds. Pp. 1–18. Cambridge, MA: MIT Press.

Cohn, Jeffrey P.
2008   Citizen Science: Can Volunteers Do Real Research? Bioscience 58(3): 192-97.
Cole, Luke W., and Sheila R. Foster
2001   From the Ground Up: Environmental Racism and the Rise of the Environmental Justice Movement. New York: New York University Press.
Collins, H. M., and Robert Evans
2002   The Third Wave of Science Studies: Studies of Expertise and Experience. Social Studies of Science 32(2): 235-96.
Collins, Harry, and Robert Evans
2007   Rethinking Expertise. Chicago: University of Chicago Press.
Collins, Jane L., Micaela di Leonardo, and Brett Williams, eds.
2008   New Landscapes of Inequality: Neoliberalism and the Erosion of Democracy in America. Santa Fe, NM: School for Advanced Research.
Cooper, Mark H.
2009   Commercialization of the University and Problem Choice by Academic Biological Scientists. Science, Technology, and Human Values 34(5): 629-53.
Corburn, Jason
2005   Street Science: Community Knowledge and Environmental Health Justice. Cambridge, MA: MIT Press.
Daston, Lorraine, and Peter Galison
2007   Objectivity. New York: Zone Books.
Daston, Lorraine, and H. Otto Sibum
2003   Introduction: Scientific Personae and Their Histories. Science in Context 16(1/2): 1-8.
Di Chiro, Giovanna
1997   Local Actions, Global Visions: Remaking Environmental Expertise. Frontiers 18(2): 203-31.
Dickinson, Janis L., Benjamin Zuckerberg, and David N. Bonter
2010.   Citizen Science as an Ecological Research Tool: Challenges and Benefits. Annual Review of Ecology, Evolution, and Systematics 41: 149-72.
Downey, Gary Lee
2005   Are Engineers Losing Control of Technology? From "Problem Solving" to "Problem Definition and Solution" in Engineering Education. Transactions of the Institute of Chemical Engineers 83(A6): 583-95.
Downey, Gary Lee, et al.
2006   The Globally Competent Engineer: Working Effectively with People Who Define Problems Differently. Journal of Engineering Education 95(2): 107-22.
Doyle, Jack
2002   Riding the Dragon: Royal Dutch Shell and the Fossil Fire. Boston: Environmental Health Fund.
Dunn, Laura
2001   Gasoline Alley Tour: SEED Coalition.
Ellis, Rebecca, and Claire Waterton
2004   Environmental Citizenship in the Making: The Participation of Volunteer Naturalists in UK Biological Recording and Biodiversity Policy. Science and Public Policy 31(2): 95-105.
Epstein, Steven
1995   The Construction of Lay Expertise: AIDS Activism and the Forging of Credibility in the Reform of Clinical Trials. Science, Technology, and Human Values 20(4): 408-37.

1996   Impure Science: AIDS, Activism, and the Politics of Knowledge. Berkeley: University of California Press.

Faulkner, Wendy

2007   "Nuts and Bolts and People": Gender-Troubled Engineering Identities. Social Studies of Science 37(3): 331-56.

Fiorino, Daniel J.

2006   The New Environmental Regulation. Cambridge, MA: MIT Press.

Fischer, Frank

1990   Technocracy and the Politics of Expertise. Newbury Park, CA: Sage.

2000   Citizens, Experts, and the Environment: The Politics of Local Knowledge. Durham, NC: Duke University Press.

Fortun, Kim

2001   Advocacy after Bhopal: Environmentalism, Disaster, New Global Orders. Chicago: University of Chicago Press.

Foster, Sheila

2002   Environmental Justice in an Era of Devolved Collaboration. In Justice and Natural Resources: Concepts, Strategies, and Applications. K. M. Mutz, G. C. Bryner, and D. S. Kenney, eds. Washington, DC: Island Press.

Fraser, Nancy

1992   Rethinking the Public Sphere: A Contribution to the Critique of Actually Existing Democracy. In Habermas and the Public Sphere. C. Calhoun, ed. Pp. 109-42. Cambridge, MA: MIT Press.

Freeman, A. Myrick

2006   Economics, Incentives, and Environmental Policy. In Environmental Policy: New Directions for the Twenty-First Century. N. J. Vig and M. E. Kraft, eds. Washington, DC: CQ Press.

Freudenberg, William R.

2005   Seeding Science, Courting Conclusions: Reexamining the Intersection of Science, Corporate Cash, and the Law. Sociological Forum 20(1): 3-33.

Frickel, Scott

2008   On Missing New Orleans: Lost Knowledge and Knowledge Gaps in an Urban Hazardscape. Environmental History 13(4): 643-50.

Frickel, Scott, Sahra Gibbon, Jeff Howard, Joanna Kepner, Gwen Ottinger, and David J. Hess

2010   Undone Science: Charting Social Movement and Civil Society Challenges to Research Agenda-Setting. Science, Technology, and Human Values 35(4): 444-73.

Funtowicz, Silvio O., and Jerome R. Ravetz

1992   Three Types of Risk Assessment and the Emergence of Post-Normal Science. In Social Theories of Risk. S. Krimsky and D. Golding, eds. Pp. 251-73. Westport, CT: Praeger.

Gieryn, Thomas F.

1999   Cultural Boundaries of Science: Credibility on the Line. Chicago: University of Chicago Press.

Givel, Michael

2007   Motivation of Chemical Industry Social Responsibility through Responsible Care. Health Policy 81: 85-92.

Gray, Leonard

2000a   Good Hope Town History. River Current Magazine, December/January 2000: 27.

2000b New Sarpy Town History. River Current Magazine, December/January 2000 : 49.

2000c Norco Town History. River Current Magazine, December/January 2000: 51.

Gray, Leonard, and Amy Szpara
2001 Orion Tank Fire Erupts. (Laplace, LA) L'Observateur, June 9: 1, 2.

Gregory, Steven
1999 Black Corona: Race and the Politics of Place in an Urban Community. Princeton, NJ:
Princeton University Press.

Grünberg, Slawomir, dir.
2002 Fenceline: A Company Town Divided. Distributed by LOGTV, Ltd.

Guarisco, Tom
2001 Activists Protest Breaux's Pollution Stance. (Baton Rouge) Louisiana State-Times/Morning
Advocate, July 28: 2B.

Guldbrandsen, Thaddeus C., and Dorothy Holland
2001 Encounters with the Super-Citizen: Neoliberalism, Environmental Activism, and the
American Heritage Rivers Initiative. Anthropological Quarterly 74(3): 124-34.

Gunningham, Neil
1995 Environment, Self-Regulation, and the Chemical Industry: Assessing Responsible Care.
Law and Policy 17(1): 57-109.

Gunningham, Neil, Robert A. Kagan, and Dorothy Thornton
2004 Social License and Environmental Protection: Why Businesses Go beyond Compliance.
Law and Social Inquiry 29(2): 307-41.

Guston, David
1999 Evaluating the First U.S. Consensus Conference: The Impact of Citizens' Panel on Telecommunications and the Future of Democracy. Science, Technology, and Human Values
24(4): 451-82.

Habermas, Jurgen
1989 [1962] The Structural Transformation of the Public Sphere: An Inquiry into a Category of
Bourgeois Society. T. Burger, transl. Cambridge, MA: MIT Press.

Hackett, Edward J., Olga Amsterdamska, Michael Lynch, and Judy Wajcman, eds.
2008 The Handbook of Science and Technology Studies. Cambridge, MA: MIT Press.

Haggerty, Julia Hobson
2007 "I'm Not a Greenie, But . . .": Environmentality, Eco-Populism and Governance in New
Zealand; Experiences from the Southland White Bait Fishery. Journal of Rural Studies 23:
222-37.

Harris, Stuart G., and Barbara L. Harper
1997 A Native American Exposure Scenario. Risk Analysis 17(6): 789-95.

Harvey, David
1989 From Managerialism to Entrepreneurialism: The Transformation of Urban Governance in
Late Capitalism. Geografiska Annaler 71(1): 3-17.

2005 A Brief History of Neoliberalism. Oxford: Oxford University Press.

Head, Rebecca A.
1995 Health-Based Standards: What Role in Environmental Justice? In Environmental Justice:
Issues, Policies, Solutions. B. Bryant, ed. Pp. 45-56. Washington, DC: Island Press.

Hess, David J.
2007 Alternative Pathways in Science and Industry: Activism, Innovation, and the Environment
in an Era of Globalization. Cambridge, MA: MIT Press.

Heynen, Nik, James McCarthy, Scott Prudham, and Paul Robbins

2007   Introduction: False Promises. *In* Neoliberal Environments: False Promises and Unnatural Consequences. N. Heynen, J. McCarthy, S. Prudham, and P. Robbins, eds. Pp. 1-21. London: Routledge.

Heynen, Nik, James McCarthy, Scott Prudham, and Paul Robbins, eds.

2007   Neoliberal Environments: False Promises and Unnatural Consequences. London: Routledge.

Hoare, Quintin, and Geoffrey Nowell Smith, eds.

1971   Selections from the Prison Notebooks of Antonio Gramsci. New York: International Publishers.

Hoffman, Andrew J.

1997   From Heresy to Dogma: An Institutional History of Corporate Environmentalism. San Francisco: New Lexington Press.

Holifield, Ryan

2007   Neoliberalism and Environmental Justice Policy. *In* Neoliberal Environments: False Promises and Unnatural Consequences. N. Heynen, J. McCarthy, S. Prudham, and P. Robbins, eds. Pp. 202-14. London: Routledge.

Holland, Dorothy, Donald M. Nonini, Catherine Lutz, Lesley Bartlett, Marla Frederick-McGlathery, Thaddeus C. Guldbrandsen, and Enrique G. Murillo

2007   Local Democracy under Siege: Activism, Public Interests, and Private Politics. New York: New York University Press.

Howard, Jennifer, Jennifer Nash, and John Ehrenfeld

1999   Industry Codes as Agents of Change: Responsible Care Adoption by U.S. Chemical Companies. Business Strategy and the Environment 8: 281-95.

Howard-Grenville, Jennifer, Jennifer Nash, and Cary Coglianese

2008   Constructing the License to Operate: Internal Factors and Their Influence on Corporate Environmental Decisions. Law and Policy 30(1): 73-107.

Infante, Peter F.

2006   The Past Suppression of Industry Knowledge of the Toxicity of Benzene to Humans and Potential Bias in Future Benzene Research. International Journal of Occupational and Environmental Health 12(3): 268-72.

Irwin, Alan

1995   Citizen Science: A Study of People, Expertise, and Sustainable Development. London: Routledge.

2001   Constructing the Scientific Citizen: Science and Democracy in the Biosciences. Public Understanding of Science 10(1): 1-18.

Irwin, Alan, and Brian Wynne

1996   Introduction. *In* Misunderstanding Science? The Public Reconstruction of Science and Technology. A. Irwin and B. Wynne, eds. Pp. 1-17. Cambridge: Cambridge University Press.

Irwin, Alan, and Brian Wynne, eds.

1996   Misunderstanding Science? The Public Reconstruction of Science and Technology. Cambridge: Cambridge University Press.

Jaffe, JoAnn, and Amy A. Quark

2006   Social Cohesion, Neoliberalism, and the Entrepreneurial Community in Rural Saskatchewan. American Behavioral Scientist 50(2): 206-25.

Jasanoff, Sheila

1990   The Fifth Branch: Science Advisors as Policymakers. Cambridge, MA: Harvard University Press.

Jessop, Bob
2002   Liberalism, Neoliberalism, and Urban Governance. Antipode 34(3): 452–72.

Johnson, Jaclyn R., and Darren Ranco
2011   Risk Assessment and Native Americans at the Cultural Crossroads: Making Better Science or Redefining Health? In Technoscience, Environmental Justice, and the Spaces Between: Transforming Expert Cultures through Grassroots Engagement. G. Ottinger and B. Cohen, eds. Pp. 179–99. Cambridge, MA: MIT Press.

Joss, Simon
1998   Danish Consensus Conferences as a Model of Participatory Technology Assessment: An Impact Study of Consensus Conferences on Danish Parliament and Danish Public Debate. Science and Public Policy 25(1): 2-22.

Kahn, Joseph
2001   Criticism and Support for Rules on Clean Air. New York Times, July 11: 13.

Karkkainen, Bradley C., Archon Fung, and Charles F. Sabel
2000   After Backyard Environmentalism: Toward a Performance-Based Regime of Environmental Regulation. American Behavioral Scientist 44(4): 692-714.

Keim, Brandon
2009   Crowdsourcing for Plants. Wired Science, May 18. http://www.wired.com/wiredscience/2009/05/foragerdata/.

King, Andrew A., and Michael J. Lenox
2000   Industry Self-Regulation without Sanctions: The Chemical Industry's Responsible Care Program. Academy of Management Journal 43(4): 698-716.

Knorr Cetina, Karin D.
1981   The Manufacture of Knowledge: An Essay on the Constructivist and Contextual Nature of Science. Oxford: Pergamon.

Kochtcheeva, Lada V.
2009   Comparative Environmental Regulation in the United States and Russia: Institutions, Flexible Instruments, and Governance. Albany: SUNY Press.

Kuehn, Robert R.
1996   The Environmental Justice Implications of Quantitative Risk Assessment. University of Illinois Law Review 38: 103-72.

Kuhn, Thomas
1996   The Structure of Scientific Revolutions. Chicago: University of Chicago Press.

Latour, Bruno
1987   Science in Action: How to Follow Scientists and Engineers through Society. Cambridge, MA: Harvard University Press.

Latour, Bruno, and Steve Woolgar
1986   Laboratory Life: The Construction of Scientific Facts. Princeton, NJ: Princeton University Press.

Lave, Rebecca, Philip Mirowski, and Samuel Randalls
2010   Introduction: STS and Neoliberal Science. Social Studies of Science 40(5): 659-75.

Law, John
1987   Technology and Heterogeneous Engineering: The Case of Portugese Expansion. In The Social Construction of Technological Systems: New Directions in the Sociology and

History of Technology. W. E. Bijker, T. P. Hughes, and T. Pinch, eds. Pp. 111-34. Cambridge, MA: MIT Press.

LeBlanc, Eric

2001   Gasoline Alley Victims Tour Stops in New Sarpy. (Boutte, LA) St. Charles Herald-Guide, August 1, 2001: A1, A3.

Lerner, Steve

2005   Diamond: A Struggle for Justice in Louisiana's Chemical Corridor. Cambridge, MA: MIT Press.

Liévanos, Raoul S., Jonathan K. London, and Julie Sze

2011   Uneven Transformations and Environmental Justice: Regulatory Science, Street Science, and Pesticide Regulation in California. In Technoscience and Environmental Justice: Expert Cultures in a Grassroots Movement. G. Ottinger and B. R. Cohen, eds. Pp. 201-28. Cambridge, MA: MIT Press.

Loopmans, Maarten P. J., Pascal de Decker, and Chris Kesteloot

2010   Social Mix and Passive Revolution: A Neo-Gramscian Analysis of the Social Mix Rhetoric in Flanders, Belgium. Housing Studies 25(2): 181-200.

Lynch, William T., and Ronald Kline

2000   Engineering Practice and Engineering Ethics. Science, Technology, and Human Values 25 (2): 195-225.

Lynn, Frances M., George Busenberg, Nevin Cohen, and Caron Chess

2000   Chemical Industry's Community Advisory Panels: What Has Been Their Impact? Environmental Science and Technology 34(10): 1881-86.

Lynn, Frances M., and Caron Chess

1994   Community Advisory Panels within the Chemical Industry: Antecedents and Issues. Business Strategy and the Environment 3(2): 92-99.

Mansbridge, Jane, with James Bohman, Simone Chambers, David Estlund, Andreas Føllesdal, Archon Fung, Cristina Lafont, Bernard Manin, and José Luis Martí

2010   The Place of Self-Interest and the Role of Power in Deliberative Democracy. Journal of Political Philosophy 18(1): 64-100.

Martin, Brian

2006   Strategies for Alternative Science. In The New Political Sociology of Science: Institutions, Networks, and Power. S. Frickel and K. Moore, eds. Pp. 272-98. Madison: University of Wisconsin Press.

McGoey, L., and E. Jackson

2009   Seroxat and the Suppression of Clinical Trial Data: Regulatory Failure and the Uses of Legal Ambiguity. Journal of Medical Ethics 35(2): 107-12.

Merton, Robert K.

1979 [1942]   The Normative Structure of Science. In The Sociology of Science: Theoretical and Empirical Investigations. R. K. Merton, ed. Pp. 267-80. Chicago: University of Chicago Press.

Michaels, David

2008   Doubt Is Their Product: How Industry's Assault on Science Threatens Your Health. Oxford: Oxford University Press.

Miller, Clark A.

2001   Challenges in the Application of Science to Global Affairs: Contingency, Trust, and Moral Order. In Changing the Atmosphere: Expert Knowledge and Environmental Governance. C. A. Miller and P. N. Edwards, eds. Pp. 247-85. Cambridge, MA: MIT Press.

Miller, Clark A., and Paul N. Edwards, eds.

2001   Changing the Atmosphere: Expert Knowledge and Environmental Governance. Cambridge, MA: MIT Press.

Moffatt, Suzanne, Peter Phillimore, E. Hudson, and D. Downey.
2000   "Impact? What Impact?" Epidemiological Research Findings in the Public Domain: A Case Study from North-East England. Social Science and Medicine 51(12): 1755-69.

Moore, Kelly, Daniel L. Kleinman, David Hess, and Scott Frickel
2011   Science and Neoliberal Globalization. Theory and Society 40(5): 505-32.

National Science Foundation
1998   Science and Engineering Indicators. Washington, DC: National Science Board.

Nguyen, Vinh-Kim, and Karine Peschard
2003   Anthropology, Inequality, and Disease: An Overview. Annual Review of Anthropology 32: 447-74.

Nieusma, Dean
2007   Challenging Knowledge Hierarchies: Working toward Sustainable Development in Sri Lanka's Energy Sector. Sustainability: Science, Practice, Policy 3(1): 32-44.

Nishizawa, Mariko, and Ortwin Renn
2006   Responding Public Demand for Assurance of Genetically Modified Crops: Case from Japan. Journal of Risk Research 9(1): 41-56.

Noble, David F.
1977   America by Design: Science, Technology, and the Rise of Corporate Capitalism. Oxford: Oxford University Press.

Oglesby, Elizabeth
2004   Corporate Citizenship? Elites, Labor, and the Geographies of Work in Guatemala. Environment and Planning D 22(4): 553-72.

Oreskes, Naomi, and Eric M. Conway
2010   Merchants of Doubt: How a Handful of Scientists Obscured the Truth on Issues from Tobacco Smoke to Global Warming. New York: Bloomsbury Press.

Organization for Economic Cooperation and Development
1997   Promoting Public Understanding of Science and Technology. Paris: Organization for Economic Cooperation and Development.

O'Rourke, Dara, and Gregg P. Macey
2003   Community Environmental Policing: Assessing New Strategies of Public Participation in Environmental Regulation. Journal of Policy Analysis and Management 22(3): 383-414.

Ottinger, Gwen
2006   Belief in "Cancer Alley": Church, Chemicals, and Community in New Sarpy, Louisiana. In Dispatches from the Field: Neophyte Ethnographers in a Changing World. A. Gardner and D. M. Hoffman, eds. Pp. 153-66. Long Grove, IL: Waveland Press.
2008   Assessing Community Advisory Panels: A Case Study from Louisiana's Chemical Corridor. In Studies in Sustainability White Paper Series. Philadelphia: Chemical Heritage Foundation.
2009   Epistemic Fencelines: Air Monitoring Instruments and Expert-Resident Boundaries. Spontaneous Generations 3(1): 55-67.
2010   Buckets of Resistance: Standards and the Effectiveness of Citizen Science. Science, Technology, and Human Values 35(2): 244-70.
2011a   Environmentally Just Technology. Environmental Justice 4(1): 81-85.
2011b   Rupturing Engineering Education: Opportunities for Transforming Expert Identities through Community-Based Projects. In Technoscience and Environmental Justice: Expert

Cultures in a Grassroots Movement. G. Ottinger and B. R. Cohen, eds. Pp. 229-48. Cambridge, MA: MIT Press.

Ottinger, Gwen, and Benjamin R. Cohen

2011   Technoscience and Environmental Justice: Expert Cultures in a Grassroots Movement. Cambridge, MA: MIT Press.

Ottinger, Gwen, and Rachel Zurer

2011   Drowning in Data. Issues in Science and Technology 27(3): 71-73, 76-77, 80-82.

Overdevest, Christine, and Brian Mayer

2008   Harnessing the Power of Information through Community Monitoring: Insights from Social Science. Texas Law Review 86: 1493-1526.

Pardo, Mary

1998   Mexican American Women Activists: Identity and Resistance in Two Los Angeles Communities. Philadelphia: Temple University Press.

Pearson, Thomas

2009   On the Trail of Genetically Modified Organisms: Environmentalism within and against Neoliberal Order. Cultural Anthropology 24(4): 712-45.

Peck, Jamie, and Adam Tickell

2002   Neoliberalizing Space. Antipode 34(3): 380-404.

Pérez, Gina M.

2008   Discipline and Citizenship: Latino/a Youth in Chicago JROTCs Programs. In New Landscapes of Inequality: Neoliberalism and the Erosion of Democracy in America. J. L. Collins, M. di Leonardo, and B. Williams, eds. Pp. 113-30. Santa Fe, NM: School for Advanced Research Press.

Phillimore, Peter, and Suzanne Moffatt

2004   "If we have wrong perceptions of our area, we cannot be surprised if others do as well": Representing Risk in Teeside's Environmental Politics. Journal of Risk Research 7(2): 171-84.

Powell, Maria, and Jim Powell

2011   Invisible People, Invisible Risks: How Scientific Assessments of Environmental Health Risks Overlook Minorities—and How Community Participation Can Make Them Visible. In Technoscience, Environmental Justice, and the Spaces Between: Transforming Expert Cultures through Grassroots Engagement. G. Ottinger and B. Cohen, eds. Pp. 149–78. Cambridge, MA: MIT Press.

Press, Daniel, and Daniel A. Mazmanian

2006   The Greening of Industry: Combining Government Regulation and Voluntary Strategies. In Environmental Policy: New Directions for the Twenty-First Century. N. J. Vig and M. E. Kraft, eds. Washington, DC: CQ Press.

Reynolds, Terry S.

1991   The Engineer in 20th-Century America. In The Engineer in America: A Historical Anthology from Technology and Culture. T. S. Reynolds, ed. Chicago: University of Chicago Press.

Riley, Donna, and A. H. Bloomgarden

2006   Learning and Service in Engineering and Global Development. International Journal for Service Learning in Engineering 2(1): 48-59.

Roberts, J. Timmons, and Melissa M. Toffolon-Weiss

2001   Chronicles from the Environmental Justice Frontline. Cambridge: Cambridge University Press.

Rose, Nikolas

1996a The Death of the Social? Re-Figuring the Territory of Government. Economy and Society 25(3): 327-56.

1996b Inventing Our Selves: Psychology, Power, and Personhood. Cambridge: Cambridge University Press.

1999 Powers of Freedom: Reframing Political Thought. Cambridge: Cambridge University Press.

2001 The Politics of Life Itself. Theory, Culture, and Society 18(6): 1–30.

Rousseau, Max

2009 Re-imagining the City Centre for the Middle Classes: Regeneration, Gentrification, and Symbolic Policies in "Loser Cities." International Journal of Urban and Regional Research 33(3): 770-88.

Rowe, Gene, and Lynn J. Frewer

2004 Evaluating Public-Participation Exercises: A Research Agenda. Science, Technology, and Human Values 29(4): 512-57.

Rowe, Gene, Roy Marsh, and Lynn J. Frewer

2004 Evaluation of a Deliberative Conference. Science, Technology, and Human Values 29(1): 88-121.

Royal Society

1985 The Public Understanding of Science. London: Royal Society.

Ryder, Paul, ed.

2006 Good Neighbor Campaign Handbook: How to Win. New York: iUniverse.

Sadler, David, and Stuart Lloyd

2009 Neo-liberalising Corporate Social Responsibility: A Political Economy of Corporate Citizenship. Geoforum 40: 613-22.

Sawyer, Suzana

2004 Crude Chronicles: Indigenous Politics, Multinational Oil, and Neoliberalism in Ecuador. Durham, NC: Duke University Press.

Schlosberg, David

1999 Environmental Justice and the New Pluralism: The Challenge of Difference for Environmentalism. Oxford: Oxford University Press.

Secord, Anne

2003 "Be what you would seem to be": Samuel Smiles, Thomas Edward, and the Making of a Working-Class Scientific Hero. Science in Context 16(1/2): 147-73.

Shamir, Ronen

2005 Mind the Gap: The Commodification of Corporate Social Responsibility. Symbolic Interaction 28(2): 229-53.

2008 The Age of Responsibilization: On Market-Embedded Morality. Economy and Society 37(1): 1-19.

2010 Capitalism, Governance, and Authority: The Case of Corporate Social Responsibility. Annual Review of Law and Social Science 6: 531-53.

Shapin, Steven

1994 A Social History of Truth: Civility and Science in Seventeenth-Century England. Chicago: University of Chicago Press.

2003 Trusting George Cheyenne: Scientific Expertise, Common Sense, and Moral Authority in Early-Eighteenth-Century Dietetic Medicine. Bulletin of the History of Medicine 77(2): 263-97.

2008   The Scientific Life: A Moral History of a Late Modern Vocation. Chicago: University of Chicago Press.

Shever, Elana

2008   Neoliberal Associations: Property, Company, and Family in the Argentine Oil Fields. American Ethnologist 35(4): 701-16.

2010   Engendering the Company: Corporate Personhood and the "Face" of an Oil Company in Metropolian Buenos Aires. PoLAR: Political and Legal Anthropology Review 33(1): 26–46.

Shrader-Frechette, Kristin

2002   Trading Jobs for Health: Ionizing Radiation, Occupational Ethics, and the Welfare Argument. Science and Engineering Ethics 8(2): 139-54.

2007   Taking Action, Saving Lives: Our Duties to Protect Environmental and Public Health. Oxford: Oxford University Press.

Sibley, Susan S.

2009   Taming Prometheus: Talk about Safety and Culture. Annual Review of Sociology 35: 341-69.

Simmons, Peter, and Brian Wynne

1993   Responsible Care: Trust, Credibility and Environmental Management. In Environmental Strategies for Industry: International Perspectives on Research Needs and Policy Implications. K. Fischer and J. Schot, eds. Pp. 201-26. Washington, DC: Island Press.

Sternberg, Mary

1996   Along the River Road: Past and Present on Louisiana's Historic Byway. Baton Rouge: Lousiana State University Press.

Stretesky, Paul B., and Michael J. Lynch

2009   Does Self-Policing Reduce Chemical Emissions? Social Science Journal 46: 459-73.

Suchman, Lucy

2000   Organizing Alignment: A Case of Bridge Building. Organization 7(2): 311-27.

Swerczek, Mary

2001   Lightning Ignites Fire at Norco Refinery. (New Orleans) Times-Picayune, June 8, 2001: A1, A14.

Swierstra, Tsjalling, and Jaap Jelsma

2006   Responsibility without Moralism in Technoscientific Design Practice. Science, Technology, and Human Values 31(3): 309-32.

Sze, Julie

2007   Noxious New York: The Racial Politics of Urban Health and Environmental Justice. Cambridge, MA: MIT Press.

Tapper, Richard

1997   Voluntary Agreements for Environmental Performance Improvement: Perspectives on the Chemical Industry's Responsible Care Programme. Business Strategy and the Environment 6: 28-92.

Tesh, Sylvia Noble

2000   Uncertain Hazards: Environmental Activists and Scientific Proof. Ithaca, NY: Cornell University Press.

Thornton, Dorothy, Robert A. Kagan, and Neil Gunningham

2009   When Social Norms and Pressures Are Not Enough: Environmental Performance in the Trucking Industry. Law and Society Review 43(2): 405-36.

Toffolon-Weiss, Melissa, and Timmons Roberts

2005   Who Wins, Who Loses? Understanding Outcomes of Environmental Injustice Struggles. *In* Power, Justice, and the Environment: A Critical Appraisal of the Environmental Justice Movement. David Naguib Pellow and Robert J. Brulle, eds. Pp. 77–90. Cambridge, MA: MIT Press.

Tong, S., and J. Olsen
2005   The Threat to Scientific Integrity in Occupational and Environmental Medicine. Occupational and Environmental Medicine 62(12): 843-46.

Traweek, Sharon
1988   Beamtimes and Lifetimes: The World of High Energy Physicists. Cambridge, MA: Harvard University Press.

Trevelyan, James
2010   Reconstructing Engineering from Practice. Engineering Studies 2(3): 175-95.

United States Environmental Protection Agency
2008   EPA's Environmental Justice Collaborative Problem-Solving Model. Washington, DC: United States Environmental Protection Agency. http://www.epa.gov/compliance/ej/resources/publications/grants/cps-manual-12-27-06.pdf.

Ward, Roger, and Janice Dickerson
2001   Community-Industry Panel (SOP). Baton Rouge, LA: Louisiana Department of Environmental Quality, Community-Industry Relations Group.

Watts, Michael
2005   Righteous Oil? Human Rights, the Oil Complex, and Corporate Social Responsibility. Annual Review of Environment and Resources 30: 373–407.

Welker, Marina
2009   "Corporate Security Begins in the Community": Mining, the Corporate Social Responsibility Industry, and Environmental Advocacy in Indonesia. Cultural Anthropology 24(1): 142-79.

Wynne, Brian
1996   May the Sheep Safely Graze? A Reflexive View of the Expert-Lay Knowledge Divide. *In* Risk, Environment, and Modernity: Towards a New Ecology. S. Lash, B. Szerszynski, and B. Wynne, eds. Pp. 44-83. London: Sage.
2003   Seasick on the Third Wave? Subverting the Hegemony of Propositionalism: Response to Collins and Evans (2002). Social Studies of Science 33(3): 401-17.

Young, Iris Marion
2001   Activist Challenges to Deliberative Democracy. Political Theory 29(5): 670-90.
2006   Responsibility and Global Justice: A Social Connection Model. Journal of Political Philosophy 23(1): 102-30.

Zavestocki, Stephen, Phil Brown, Meadow Linder, Sabrina McCormick, and Brian Mayer
2002   Science, Policy, Activism, and War: Defining the Health of Gulf War Veterans. Science, Technology, and Human Values 27(2): 171-205.

Zyglidopoulos, Stelios C.
2002   The Social and Environmental Responsibilities of Multinationals: Evidence from the Brent Spar Case. Journal of Business Ethics 36(1/2): 141-51.

ABOUT THE AUTHOR

Gwen Ottinger is Assistant Professor in Interdisciplinary Arts and Sciences at the University of Washington–Bothell, where she teaches in the Science, Technology, and Society and Environmental Studies programs. She is coeditor of *Technoscience and Environmental Justice: Expert Cultures in a Grassroots Movement.*

Made in the USA
San Bernardino, CA
27 January 2018